DAVID H. COWARDIN

ICS STRATEGY

BY

DAVID H. COWARDIN

TABLE OF CONTENTS

ABOUT THE AUTHOR

David H. Cowardin has worked in fire, emergency medical, and disaster management fields for more than 40 years. He has spent more than 35 years as an officer as well as over 30 years as a training officer and department head. He has served in organizations of all types and sizes from those that were all-career to ones that were mostly volunteer. The agencies have had as few as one station and as many as 40, with staffing ranging in size from 24 personnel to over 1,000. He has directed operations in heavy industrial areas, residential communities, wildland/urban interface areas, on airports, etc.

Dave has an Associates Degree from Rancho Santiago College in Fire Administration. He also holds a Bachelors Degree in Fire Science, a Masters Degree in Fire Administration, and a Doctorate in Fire Education from Western States University. A training officer for one of the original seven partner agencies under ICS-FIRESCOPE he has developed many Incident Command System (ICS) programs. He has written numerous articles on management and other related subjects, in addition to authoring two other books.

FORWARD

All officers must be taught to manage emergencies. Yet, the first decision officers make on an emergency is to manage or not to manage. We need to recognize that action is what causes emergencies to be abated. The management system is the tool used to facilitate action. Therefore, the first officer must be given the ability to initiate action or take command. The rule to follow is:

"When your hands on involvement will have a significant impact on the outcome, take action. If one more set of hands will have little or no impact, take command."

This book presents a single management system for you to use when taking command of an Incident Command System (ICS) operation. The content is based on sound management and command principles. But more importantly, it has been refined through actual utilization by career and volunteer personnel on all types and sizes of incidents.

An attempt was made to keep this text from becoming too lengthy. Size of the book was not an issue, but the body of knowledge it contains was intended to surpass all others. The book's goal is to improve your abilities to deal with day-to-day, as well as major situations, using a strategy that is proven to work with ICS.

PROLOGUE

ICS Strategy was developed over a 25-year period. The California Department of Forestry and Fire Protection (CDF) developed the original concept for managing initial attack on wildland fires. In the early 1970's this format was expanded by the author for municipal use in dealing with structure fire management, aircraft emergencies, disaster preparedness, etc. This book represents a considerable refinement of this early system. The most significant change in ICS Strategy is that the management system has been designed specifically to work with the Incident Command System (ICS).

This book answers the need for a day-to-day management system that will work with ICS. This management system will improve any officer's ability to deal with any size or type of emergency. The text is not another ICS book; rather it presents a single management method that is proven to work with ICS. ICS STRATEGY will work with or without ICS, but is designed to enhance the ICS process.

CHAPTER I

EMERGENCY MANAGEMENT SYSTEMS

INTRODUCTION

Every person in charge of an incident - - whether career, elected, volunteer, or acting - - needs a method by which direction can be given to emergency forces. It is one thing to manage an office or large corporation, but quite another to deal with injury, as well as loss of life and property. Management in today's emergency fields realistically involves two distinctively different areas of responsibility - - day-to-day management and incident management. The fundamental beginnings of both types of management are the same, and were defined and perfected during the Industrial Revolution. There have been few changes in the basic approaches since that time. Therefore, many other theories used to manage are deeply embedded in our society and respective emergency fields.

Management techniques used in most emergency fields have been influenced by what is termed "modern management," and a combination of what existed in the past or traditional approaches to emergency management. Most textbooks on emergency management tend to deal in idea-logic approaches. They often present different philosophies or theories, and let the reader extract the portions that seem to fit a particular situation or organization.

Managing a large company or organization on a day-to-day basis can be fairly simple in comparison to managing an emergency. The complexities of emergencies, number of resources, and short time periods often tax even experienced managers. This is especially true when one is faced with the realization that not everything can be saved. Many individuals with diversified backgrounds, including educational and management experience, fail when faced with an emergency situation. Emergency management is a field unlike any other - - calling for special skills, training, and abilities.

Generally, personnel from the emergency fields, such as fire and law enforcement, are more suited to be managers during incidents of a more significant nature. Usually the larger the emergency agency, the better trained the individuals in positions of authority are in dealing with the problems that may be encountered. The reasons for this are simple: usually the larger agencies experience more call volume and thus more opportunities to manage significant incidents. However, small agencies tend to utilize their resources better because of necessity. All organizations can improve the effectiveness of their managers by using a sound management system.

Managers must be taught to deal with emergencies having a major impact in the same manner they deal with day-to-day incidents; otherwise a breakdown in the management system is likely to occur when the situation reaches a critical stage. This type of breakdown is exemplified by confusion, lack of direction, lack of organization, and poor use of existing resources.

Most large-scale emergency operations start small and expand. This means that managers should handle every incident as though it will expand into a major emergency. This is a basic concept behind ICS and ICS Strategy.

The advantage of ICS STRATEGY is that this system allows individual managers to adjust quickly to a variety of problems because the concepts remain consistent, easy to apply and very effective. The type and size of the emergency has little, if any, impact. Using ICS STRATEGY, large and small agencies can maximize the use of resources, improve safety, and enhance communication.

ICS Strategy will provide individuals with the ability to manage an emergency situation even in the face of uncontrollable devastation and undetermined behavior. This is accomplished by using emergency management techniques that regulate resources in such a manner that even though the incident is escalating, guidance and direction is being provided to meet the new demands.

There are many problems that can, and do occur, during emergencies; however, none is more apparent than the failure to manage effectively or apply sound management skills. The difference between a good agency and a poor one in relation to emergency effectiveness is not equipment or personnel, but the quality of management. Management is the cornerstone of excellence. An agency that is not capable of fully utilizing its resources can be ineffective no matter how large it may be. None of us have sufficient resources or brilliant personnel to overcome poor management in the face of a mass disaster.

MANAGEMENT SYSTEMS

Early civilizations recognized the need for leadership during an emergency situation. The Romans were one of the first civilizations to utilize the principles of a senior officer making major strategic decisions when dealing with emergencies. They organized the first known Fire Brigade. The leaders chosen for early emergency management were generally military officers. The command systems developed were primarily based on military principles. Command was broken down into manageable units; yet, the supervisors in charge rarely delegated much, if any, of their authority. In the 1940's and early 1950's, changes began to occur in emergency management systems.

Lloyd Layman

Lloyd Layman developed a process he termed "size-up." Layman taught emergency managers to evaluate an incident's potential through size-up and then develop a "plan of operation." The only weakness in Lloyd Layman's system compared to today's teaching was that a "plan of operation" was just a part of size-up. In today's era we should recognize that size-up and planning are separate management functions. The first three elements of Lloyd Layman's Size-up System (Facts, Probabilities, and Own Situation) are as viable today as the day they were written. The difference is that officers must be taught to distinguish between size-up and planning as part of the overall management of emergencies.

Management By Objectives

Most emergency agencies are making changes from past management methods, systems or styles. Many agencies are adopting Management By Objectives (MBO), or some version thereof for day-to-day operations. We have become very goal and objective orientated in the way we conduct our business. Most organizations have at least looked at MBO type systems for management of emergencies.

I, too looked at MBO as an emergency management tool. What I found was that MBO can easily overload a manager during an emergency. Time is usually very limited, and during the incident's progression a number of problems continue to surface for which objectives (partial or complete) must be developed. Even an experienced MBO manager can become overwhelmed because normal tracking devices become useless in a rapidly extending or multi-faceted situation. Generally an MBO manager never really defines what is to be accomplished by the personnel committed to the incident during the emergency. Instead, the manager spends most of one's time trying to keep track of what has been established as goals and objectives; so much time, that one begins to lose sight of the problem at hand and what is transpiring with other aspects of the problem.

Incident Management Cycle

Emergency management is much more than size-up, a plan of operation, or establishment of objectives. Emergency management is a cycle involving planning, organizing, controlling, communicating, and analyzing. What I have termed the Incident Management Cycle is at the heart of our job as Incident Commanders. Our focus has to be on improving our skills in managing plan development, establishing the organization, controlling, communicating, and analyzing the results, both short- and long-term. If we cannot improve in these areas, many of us will continue to muddle our way through emergencies allowing the incident to manage us. In the process we will waste resources that are becoming scarcer all the time, and add to the confusion of the emergency we are sent to control. If we are able to improve in the incident management cycle areas, incident command can be made simple and safer to everyone involved. To be truly effective a good emergency manager must concentrate on these management principles.

PLANNING
ORGANIZING
CONTROLLING
COMMUNICATING
ANALYZING
THE
INCIDENT
MANAGEMENT
CYCLE

ICS Strategy

The Incident Management Cycle is the foundation upon which ICS STRATEGY is built. ICS STRATEGY uses the Incident Management Cycle, combined with specific objectives, in order to enable an emergency manager to control a number of activities within a short period of time. All personnel, especially those supervising the work of others regardless of their assignment, must have sight of the objective to be reached within the time frames established by the Incident Commander. In the following chapters, we shall spend the majority of this text explaining the interrelationships of the Incident Management Cycle with ICS STRATEGY. Each area effects success of ICS Strategy and cannot be successful in managing emergencies without this knowledge.

NOTES:

NOTES:

CHAPTER II

PLANNING

During an emergency, the emergency manager must focus attention on a number of functions. None is more important than Planning. Planning is the cornerstone of ICS Strategy and excellence during an emergency. Regardless of size, no agency can be effective on any incident without a plan. A plan greatly influences the ability of personnel, the control of activities, and the quality of overall performance. An organizational system will not stand alone and must be supported by a plan.

How well we perform as emergency managers depends on what elements we use in planning and what our plan intends to accomplish. A plan for most incidents is not a detailed action plan such as taught in traditional ICS classes; it is, however, an action plan.

The definition of emergency action planning is "a process, method or procedure, both strategical and tactical, which provides unified direction for resources on an incident."

RULES FOR ACTION PLANNING

There are combinations of eight strategical and tactical rules concerning emergency ICS STRATEGY action planning that an incident manager should use to govern their plan development.

Rule 1 - An Action Plan Should Be Specific

A plan does not create doubt in the minds of the personnel involved on an incident. The plan should have laser-like qualities; identifying actions to be taken and pinpointing the intended results in non-ambiguous terms.

Good ICS STRATEGY action plans have written objectives defined in one short statement. The activities that take place because of planning can be clearly identified as moving the resources toward accomplishing those objectives.

Rule 2 - An Action Plan Is Commitment To Action

Objectives defined in an action plan are commitments to action. A commitment to action cannot and must not be taken without first considering the resources that are available. A manager can develop all the grandest plans; however, if sufficient resources are not available, the process will be a waste of time. The soundness of an action plan is based on being able to identify how the available resources enroute or requested will be deployed to meet the stated commitment.

Rule 3 - Action Plans Cause Change

An action plan is designed to cause change. An action plan is not necessary if one wishes to let nature take its own course.

Should an emergency manager design an action plan to support existing conditions, changes inevitably will occur in the situation that the plan is designed to maintain. Regardless of the type of incident, an action plan should cause the emergency to be abated, confined, or controlled; thus causing change.

In order to plan, one need to be able to predict the behavior or possible behavior of the elements involved. Without such ability, the action plan and the end product produces vague objectives and uncertain results.

Rule 4 - Action Planning Causes Improvement

A good action plan will cause change; therefore, one measure of the effectiveness of a plan is visualized in the type of change it causes.

Generally, an action plan does not call for letting the emergency grow until it reaches a certain point, size or stage. An action plan should start the commitment of resources long before the incident reaches ultimate limits. The emphasis should be on improving data collection, conditions, resource utilization, or whatever is appropriate depending on the encountered type of emergency or situation.

Rule 5 - Action Plans Are Integrating At All Levels

An effective action plan calls for coordination between individuals, units, departments, or agencies involved in accomplishing the plan. A well-constructed or defined action plan always integrates the different components or operations so they become mutually reinforcing.

As the emergency becomes larger or more spread out, integrated planning becomes more important. For example, units on a major incident may become physically separated from each other by considerable distance; therefore, supervisors could be required to formulate a plan for an individual area. These plans must integrate with other areas, as well as the overall

Incident Manager's plan. The overall plan cannot stand alone without this integration unless it is infallible.

Rule 6 - Action Planning and Risks or Values

One of the most important considerations in action planning are the values involved. Risks to emergency personnel must be given first priority. No action plan can be implemented when the potential loss of life is intolerable either to the agency or public we are attempting to protect.

Rule 7 - Action Planning Is Based On Knowledge

Every manager must recognize that the most vital element in action planning is knowledge. Without knowledge, action planning becomes an exercise in speculation. An emergency manager must not only have knowledge upon which to base action plans, but also knowledge of how to use it in the planning process.

Rule 8 - Action Planning Must Have Impact At The Operational Level

The last, but perhaps most important, rule of planning is that an action plan must have impact at the operational or tactical level.

Personnel performing the operations that will lead to the organization reaching its objective need to know how their actions are contributing to the plan. They should be aware of how the actions of others are coordinating with their efforts. Because of this each supervising officer of an operational area must know the strategic objectives of the organization.

The Incident Manager needs to see that objectives set in the action plan have impact at the action level and changes are occurring in the emergency because of this planning.

Planning is not a static process. Once it starts there needs to be a continual evaluation of new information and re-evaluation of the plan based on the information and the actions that have already begun or are slated to start in the near future. This may or may not affect the refinement of the action plan, or it could cause development of a totally new approach.

DEVELOPING THE SIMPLE ACTION PLAN

The essence of ICS Strategy is the "Action Plan." The action plan identifies the strategic objective and tactical operations in a measurable and time-related manner. It also sets the stage for organizing, directing, communicating and controlling activities. No person can be expected to manage without a plan, be it ever so simple. To develop an action plan, a manager must have some knowledge of the situation. Once information has been gathered and evaluated about the situation, the person in charge is ready to develop an action plan. A good action plan should have three (3) very distinct elements that are:

1. Strategic Objective(s)
2. Time Frame
3. Tactical Tasks or Jobs

Strategic Objective

Many supervisors spend very little, if any, time considering a strategic objective because they do not feel many short-term emergencies require identification of what is to be accomplished. The problem with this theory is that planning has to become a conditioned response on the manager's part. Failure to identify an objective on each incident leads to non-use of the plan development even when there is a need. Another repercussion is subordinates need to learn that strategic objective provides direction. If a manager does not provide an objective on simple problems, subordinates will not become conditioned to working with solid direction. This, in turn, will only compound emergency management problems on larger incidents.

A strategic objective is a statement by the person in charge of an incident as to what is to be accomplished by all personnel under the manager's direction and control. To be effective a strategic objective for an emergency must have the following qualities:

a. Identifiable
b. Measurable
c. Limiting In Scope
d. Possible To Accomplish

Identifiable Objectives

A strategic objective must be clearly stated and recognized by subordinates; otherwise, there will be a natural tendency to second-guess the Incident Manager as to what is to be accomplished. On minor incidents, second-guessing can be fairly accurate; but when a large scale or complex situation is encountered, second-guessing could miss by a considerable margin. This will lead to costly errors, loss of time, wasted resources, poor coordination, excessive communication, and possible injury to personnel. Most strategic objectives in simple incidents would follow along these lines:

Disaster management - prevent deterioration of the east levy; evaluate damage on the west side.

Emergency medical - stabilize victims; extricate victims.

Fire - confine to "x" numbers of rooms and the attic; confine to the office area; confine to the second floor; etc.

Law enforcement - control crowds between C & D Streets; evacuate the south end.

When an Incident Commander provides a strategic objective on relatively simple emergencies, they are conditioning personnel to have a sense of direction before they start any operation. When personnel become accustomed to this procedure and find out how much smoother incidents progress under this type of direction, an Incident Commanders may find individuals who normally work for them, actually asking for a strategic objective. This also helps the manager, and when a major emergency occurs the effects of this type of on-the-job training may be immeasurable.

There may be more than one strategic objective; but this is not usually the case because a strategic objective is the desired result that all personnel on the incident are committed to complete. If more than one strategic objective is provided, then there becomes a progression of accomplishments. For example:

Disaster Management

On an incident such as an earthquake, there may be a progression through several strategic objectives. The first may be damage assessment; the second, establishment of relief centers and staging areas; the third heavy rescue operations. A flood problem might follow these lines: Objective #1 evacuation, #2 water containment and #3 rescue operations.

Emergency Medical

In a multi-injury car accident the incident strategic objective may just to stabilize the victims; at the same time people may be stabilizing the vehicle and providing protecting against fire, but the most important action is the stabilizing of the victim's life threatening injuries. The next strategic objective may be extraction; and finally clean up at the scene.

Fire

Fire in a four-story hotel/tenant-type building that starts small and progresses through several strategic objectives. Upon arrival of the first fire company, the fire is located in a lower floor area near an elevator shaft. Strategic objective #1 would be to confine the fire to origin and stop horizontal spread to the elevator. Initial activities start in order to accomplish this goal; however, the fire spreads to the elevator shaft and upwards to the floors above. A second strategic objective is established which is to rescue all occupants. Additional equipment and personnel are requested, and numerous activities take place including application of water on the fire. But every action is primarily to facilitate rescue or check fire extension until rescue is completed. When strategic objective #2 is accomplished, a third strategic objective is established to confine the fire to one building.

Law Enforcement

Riot situation strategic objective progression: seal off the area; part of this might be to prevent unauthorized entry into the area. Once the area is sealed the next strategic objective may be to clear and control the streets. Streets may have to be controlled one at a time and other activities such as firefighting may have to take place, but the strategic objective is still to clear and control. The final strategic objective may be to maintain order.

As a general rule if more than one objective is being accomplished at the same time they most likely are not strategic objectives, but tactical operations. Sometimes these tactical operations carry over from one strategic objective to the next.

Measurable

For many individuals in charge of a situation, one of the most difficult concepts to comprehend is that a strategic objective must be measurable. In order to establish a measurable strategic objective, emergency managers need to think in terms of objectives that are measurable in relation to the total problem. The total problem might be to "put out the fire," "save lives," "save property," "stop flooding," etc.; however, these goals are not strategic objectives and do little, if anything, to tell a person how things are progressing unless a disaster occurs.

A measurable strategic objective is a statement that provides supervisors and subordinates with a yardstick for which to measure progress or lack of progress. The examples from "identifiable objectives" above are also measurable. They are not goal statements such as "put out the fire." Instead of stating "Prevent Flooding," one might state to "Stop Deterioration of the East Levy;" instead of "Save Lives," it might be better stated in a measurable term such as "Evacuate the Area Between X and Z

Streets." With statements of this nature if a problem occurs, you are going to be advised much earlier in the situation that the strategic objective can not be accomplished. In some cases when you are advised that the strategic objective cannot be obtained, you may not be able to do anything about the situation; but you will be ready to realign your objective sooner without committing additional forces to something that is not possible to obtain. Hopefully, feedback will be received soon enough for you to provide additional assistance or take the appropriate steps to enable you to meet the objective.

Limiting in Scope

The scope of any situation or job specification identifies what is to be accomplished. For example, today many teachers identify what they expect students to learn during a semester or year. Going one step further, most large companies (as well as the emergency service fields) have job specification sheets that tell a prospective employee what the job duties are and what is expected of them.

Limiting in scope relates to what we are attempting to accomplish. It makes a large problem into a manageable unit. It further identifies the limit of what we expect. The more limiting the scope of the strategic objective, the sooner we can measure progress or the lack of it. For example, in a fire situation when you arrive on scene of a working fire, the fire has consumed a portion of something. We still have to position companies, place equipment etc. The question is where will the fire be when we have all the resources in place to control the problem? The more limiting the scope (tighter restraints) the sooner we will be able to measure the impact of our actions. For example, in a high tide/flooding situation the limit of the strategic objective scope would be "Hold the damage to a given number of homes." What we have said is that we are going to concentrate our efforts on a small part of the total problem - save what we can and not use up or waste our available resources on other homes that are impossible to save. Using this approach, limiting the strategic objective become the game plan for the emergency -- limiting what is to be accomplished by those resources committed or to be committed.

Possible to Accomplish

Last, but certainly not least, a strategic objective must be possible to accomplish. Some managers operate on a day-to-day basis asking more than what they can reasonably expect out of their employees. They also know that they want considerably less than what they are asking. During emergencies this can become self-defeating in that employees either exert too much energy attempting to accomplish the impossible, or realize the futility of the situation and become complacent toward the strategic objective.

Managers should use the approach of setting strategic objectives that can be accomplished, even if by doing so they then expand the limits or scope of the incident to a slightly larger area. For example, a fire is in a large warehouse where we may not be able to determine the total extent of the problem. In this case we set the strategic objective to confine the fire to a wing or large area. Then as we assign and deploy companies they will move through the large area until they find the involved area and confine the fire. A subordinate officer in this situation will also send back information on the extent of the problem that will help coordinate the other operational forces. This is preferable to having personnel start attacking at a midpoint of a problem; generally it is preferred to start at the outer limits and work toward the center. Using this approach, the strategic objective tends to provide more positive direction and simplifies management of the situation.

TIME FRAME

The next element of an action plan is the time frame. A time frame is the anticipated amount of time necessary to accomplish the strategic objective. Time can be either friend or foe. As an emergency manager, time is a tool for you to use. Certainly most emergency personnel realize that in life-threatening situations, seconds save lives. Likewise, the success of many strategic objectives hinges on time. Most emergency actions are centered on very short periods of time in relation to the total number of activities that have to be accomplished. Although a strategic objective may be stated in a measurable and limiting manner, this is seldom sufficient when managers must combine all actions into one plan. This is because of the different backgrounds that officers have and the fact they may be viewing the incident from a different perspective. Therefore, a strategic objective for a plan may not offer sufficient guides to subordinates; what one person may envision taking 45 minutes to accomplish may have to be accomplished in 20 minutes to achieve the strategic objective. Thus, time becomes an important element in planning.

The problems most managers have in developing an effective plan is judging and setting of realistic time frames. Recognizing the problem is one thing, but estimating its potential in a time-related manner is yet another. The most effective method of setting time frames for a plan is thorough knowledge of history. The dictionary generally defines history as a record of, or dealing with, past events. But to the person faced with an emergency, it means much more. The manager has to consider not only similar past events, but the variations that have occurred.

Past History

Experience is our strongest teacher as to what may occur in the future. Past history or the study of incidents that have taken place under similar conditions gives us a good database from which to base decisions when in command.

The problem is that many of us do not have sufficient years and/or exposure to the types of problems we might encounter. Therefore, our database lacks depth when an incident develops. To strengthen our status, we must do more than read accounts of past emergencies. It is critical that senior personnel be taught to share the wealth of knowledge they have acquired. Even retired personnel who know of past history should not be overlooked during major situations. It is a wise command officer who learns to draw on and cultivate the knowledge of others during and before an incident occurs.

Future Forecasting

The future is tied to past history and recent incident development. To become good at future forecasting, one must become an incident behavior analyst; not in the textbook sense, but by watching incidents progress and analyzing what has occurred. Computers can help in this process; but they are limited and unable to anticipate the unknown. Computers can simulate very accurate projection in some situations based on facts; but so much of a good manager's abilities are based on a sixth sense that enables the officer based on numerous factors beyond weather, building construction or topography, action taken, etc., to predict what will, could or might take place as the incident progresses. As a situation develops, we should be able to estimate what the predicted or future behavior might be minutes or hours from the present situation.

How much knowledge of history does an Incident Manager need to establish time frames? We could never have enough to know all the answers in order to set absolute times. But we should have sufficient knowledge to be able to minimize the size and number of problems; a good rule of thumb for establishing time frames is to look at a situation on 15-minute intervals. Times to accomplish a strategic objective would then be 15, 30, 45, 60 minutes, etc. Most small incidents are controlled within the first 30-45 minutes; some in 15 minutes or less. On large situations such as haz mat, hostage, floods, etc. attainment of the strategic objective will be based on obtaining and deploying additional resources; therefore, the time frame should be established using these factors as the primary criteria. Generally time in these situations evolves based on hours.

TACTICAL TASKS OR JOBS

The final element in the development of an action plan is the establishment of the tactical tasks or jobs. The two previous elements, strategic objective and time, are intended to provide a central focus for all emergency personnel as well as the Incident Commander. The tactical tasks or jobs relate to the actions that are necessary to accomplish the strategic objective within the time established. Tactical tasks or jobs are the steps, procedures, or operations that will provide for the attainment of the strategic objective.

The tactical tasks or jobs are listed as part of planning in order of importance. When identifying the steps that should take place, the manager must ask a simple question: What is the most important step or action in order to accomplish the strategic objective? This is where many supervisors have difficulty. They inherently turn to events such as "rescue" or "save/protect lives" regardless of the situation. This occurs because of the mental and ideological importance placed upon these elements in the emergency fields; however, this is not the method that should be used to establish the first tactical task or job.

A strategic objective is the primary statement as to what the entire emergency forces are being committed to accomplish and establishes a priority that is completed before moving to another strategic objective, if necessary. Unlike a strategic objective, the tactical tasks or jobs may be carried out at the same time depending on the resources that are available. For example, one group may be starting attack functions as another begins evacuation. On a larger scale, numerous events may be taking place at the same time but the direction each is taking leads toward accomplishment of the stated strategic objective within the time provided.

Tactical Tasks

A tactical operation is an assignment or grouping of jobs. It is not broken down into specific or individual jobs, but is more general in nature with a defining quality.

The first tactical task to take place should be that action which will be the most instrumental in reaching the strategic objective within the given time limits. Take for example a fire in a small shopping center; the strategic objective might be to confine the fire to the center suite and attic within 15 minutes. The first tactical task that would have the greatest impact might be ventilation; or in another situation water on the fire; yet, in another situation it could be the stopping of spread to one side or the other.

Another example of tactical tasks is as follows: during a major flood situation, the strategic objective is the evacuation of a 20-block area. The first event would identify the area with the most critical need and then others would be listed in importance of priority. A hazardous material incident might follow this line; the strategic objective is to contain the material at the plant; the first tactical operation might be downwind evacuation, followed by diking, leak control, etc. Although not stated in either example above, the ideological importance of saving lives is accomplished by taking action to curtail the threat to lives. In some cases because of number of lives at risk, the strategic objective could be the rescue of occupants; in this type of situation the priority placed on tactical tasks would be based on using the resources where the most lives could be saved with the least amount of effort.

How many tactical tasks will there be? This depends entirely on the size and scope of the emergency operation. Most simple emergencies would tend to be in the range of four to ten tactical operations. Examples are as follows:

Emergency Medical Tactical Tasks	Fire Tactical Tasks
Victim assessment	Fire attack
Victim stabilization	Ventilation
Vehicle stabilization	Primary search
Fire Protection	Control of utilities
Extrication	Exposure protection
	Salvage

Hostage Situation Tactical Tasks	Haz-Mat Tactical Tasks
Secure perimeter	Identify product(s)
Evacuate area	Establish evacuation area
Establish contact with interior	Evacuate
Establish sharp shooter positions	Establish zones
Provide crowd control	Stabilize product(s)
Cut electrical services	Package and remove product(s)

A tactical task should be as detailed as possible depending on how critical the operation is in accomplishing the strategic objective and how well trained your personnel are on the type of situation encountered. Managers must look at this problem in a way that answers the basic question of whether it is more important how the job is performed or that it is accomplished within the time required to meet the strategic objective.

Tactical Jobs

The difference between a tactical task and a job is that with the assignment of a job the Incident

Commander tells a subordinate officer "HOW" to do the work. Generally an emergency manager using more tactical jobs than tasks becomes overwhelmed with small details often losing sight of the strategic objective. A good rule of thumb is that the larger the emergency, the less an Incident Commander should tell personnel how to do an individual job. Examples of jobs and less detailed tasks follow:

Emergency Medical	Fire
Job - Place victim on backboard	Job - 1-3/4 inch attack lines at the front
Task - Package victim for transport	Task - attack at the front.

Law Enforcement	Disaster Management
Job - Force entry at the left rear door	Job - Contact radio and TV stations
Task - gain rear access	Task - Provide public information

Both tactical tasks and jobs need to follow a progression. The first tactical task or job is that action which is necessary or will have the greatest impact on attaining the strategic objective within the time frame. The second tactical task or jobs is the next most important action to attain the strategic objective; followed by the third, fourth, fifth, and so on.

Some of the tactical tasks or jobs may not be necessary to obtain the strategic objective, but they provide for safety, reduce the potential for loss of life, and control damages. Actions like search, control of utilities, and evacuation fall into this category

The tactical tasks or jobs are the steps necessary to accomplish the strategic objective. However, the assignment of a tactical objective may cause the subordinate officer to manage a number of tactical jobs. By stating a strategic objective the Incident Commander has provided guidance for all personnel. As tasks are assigned to subordinate supervisors, they should develop the jobs, within the scope of the Incident Commander's strategic objective to accomplish that operation. Ideally, the supervisor given a task would then assign jobs to others, and concentrate on managing the operation as a segment of the total plan.

Examples of Action Plans

The strategic objective, time and tactical tasks form the action plan. Depending on you own capabilities, the training of personnel, experience levels, and size as well as type of incident an action plan may be specific, general or a combination of both. Looking at some different types of incidents the following could be examples of action plans with either tactical tasks or jobs.

Disaster Management Incident

Strategic Objective	= Evacuate Downtown
Time	= Three Hours
Tactical Tasks	Tactical Jobs
Activate relocation centers	Establish "B" St. relocation center
Establish safe routes	Establish "Elm" St. relocation center
Provide PSA's	Hwy 12 is the primary safe route
Set up assistance centers	1st St. is the secondary safe route
Provide traffic control	3rd. Street is the backup safe route
	Contact radio and TV stations
	Set up park assistance center
	Set up city hall assistance center
	Traffic control Hwy 12 and 1st St.
	Officer traffic control Elm and 5th St.

Hazardous Materials Incident

Strategic Objective	= Contain Spill to Storage Facility
Time	= One hour
Tactical Tasks	Tactical Jobs
Set-up staging areas	Staging at the high school
Evacuate victims	Incident staging behind building
Control leak source	Rescue known victims
Control run-off	Evacuate the facility
Setup decontamination	Evacuate down wind
area	Shut down "A" valve
	Shut down "B" valve
	Shut down pumps
	Build dike on 1st Street side
	Build secondary dike at 2nd St.
	Setup hot-zone deacon by pit
	Setup warm zone deacon at office

Emergency Medical Incident.

Strategic Objective	= Extricate 2 Victims
Time	=30 Minutes
Tactical Tasks	Tactical Jobs
Provide fire protection	30# dry chemical extinguisher
Stabilize victim's	1-1/2 hose line at rear
Force entry	Control bleeding
Package victim's	Maintain airways
Transport	Establish IV on victim "B"
	Use high pressure spreader right door
	Set-up helicopter spot
	Remove left passenger door
	Place victim "A" on half board
	Place victim "B" on back board
	Victim "A" to ground ambulance
	Victim "B" to air ambulance

Fire Incident

Strategic Objective	= Contain Fire to Three Rooms
Time	= 30 Minutes
Tactical Tasks	Tactical Jobs
Ventilation	Cut hole in roof over center of building
Attack	2-1/2 on main fire
Primary Search	1-1/2 to protect shop area
Utilities	Search 1st floor
Salvage	Search 2nd floor
	Search basement
	Shut off gas
	Shut off electricity
	Control water run off
	Cover furnishing on 1st floor
	Remove files from basement

Law Enforcement Incident

Strategic Objective	= Hold Rioting Between 1st & 3rd Ave.
Time	= Two Hours
Tactical Tasks	Tactical Jobs
Setup staging areas	Officers to stage at Central Park
Blockade at 1st Ave	Riot team stage at Sears parking lot
Blockade at 3rd Ave	Cars 3 & 7 blockade Main @ 3rd
Deploy riot team	Cars 2, 5, & 6 blockade Main @ 1st
Setup holding area	Car 10 patrol "A" St. between 1st & 3rd
Provide jail transport	Car 9 patrol Elm between 1st & 3rd
	Set-up holding area at City Hall
	Car 20 start sweep at 3rd and work north
	Car 15 on secondary sweep after 1st
	Second sweep to hold resisters for 1st
	Move resisters to holding area
	Use City bus to move resisters to jail

Note the use of tactical tasks leaves the development of jobs to the individuals who will be assigned the tactical task. This allows the Incident Commander to concentrate on managing the strategic objective and not get caught up in the detail of telling people how to do their job. This is very important on larger incidents.

INCIDENT PLANNING SHEETS

The Incident Commander needs to keep track of the plan (strategic objective, time frame, and tactical tasks or jobs) by writing down guides on a sheet of paper. Even better is the use of a simple incident planning sheet (see figure 2-1). This Incident Planning Sheet provides a space for writing down the strategic objective, time frame, and tactical operations. In addition, there is space for the units enroute (unit box), in staging (a check box identified with an "S"), and a place to note the assignment. The space at the bottom of the page is used to draw a quick sketch of the incident so positioning can be noted for quick reference. Also, check boxes can be added as reminders of what agencies to contact for additional support.

Of course you can develop your own sheet to meet your agency's specific needs. The only advice we have to offer in this regard is to keep it simple. ICS strategy is designed to help you, the Incident Commander, focus on what is important. So should any sheet you use to manage an incident.

INCIDENT MANAGEMENT SHEET			
STRATEGIC OBJECTIVE(S)	**UNIT**	**S**	**ASSIGN**
TIME FRAME			
TACTICAL ASSIGNMENTS, TASKS, OR JOB			

INCIDENT DRAWING

POWER		GAS		TRAFFIC		PRIMARY		SECONDARY

FIGURE 2-1 - INCIDENT MANAGEMENT SHEET

NOTES:

CHAPTER III

ORGANIZATION AND ICS

ICS ORGANIZATIONAL STRUCTURES

In this chapter we will discuss the Incident Command System (ICS) and its practical application to most emergencies. We will not deal in detail with an element, only application to day-to-day situations. If you need more complete information contact your State or Federal Emergency Management Agency.

When managers blend people, knowledge, and organizational relationships together they form an organization. An organizational structure is composed of interrelated positions that are arranged by a

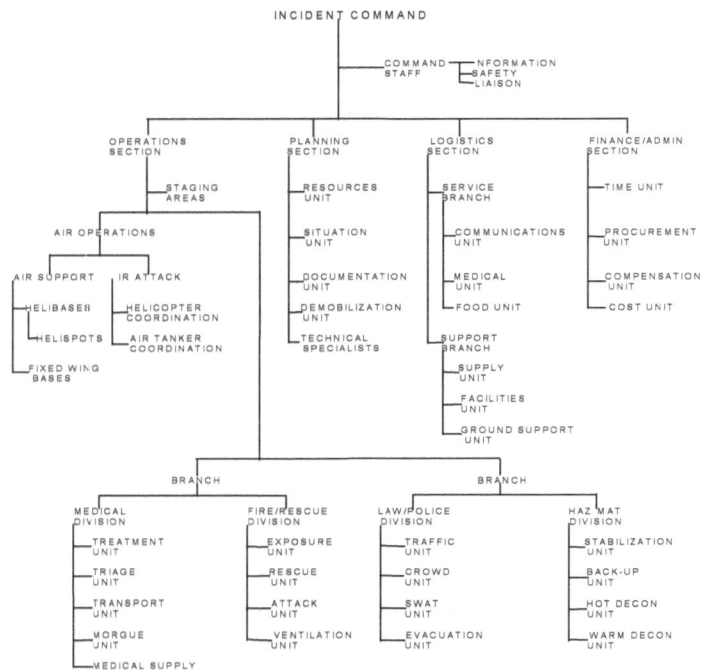

FIGURE 3-1 MULTI-FUNCTION ICS ORGANIZATION

manager in order to achieve objectives in an efficient and timely manner. ICS is a line-and-staff-type organization that can be modified to meet any type of emergency incident organizational needs. Figure 3-1 is an example of a multi-faceted ICS organization covering each of the four major emergency fields: medical, fire/rescue, law, and hazardous materials.

The purpose of an ICS line-and-staff organization is to provide needed assistance to line or operations supervisors. ICS staff personnel handle specific

46

advisory and support functions that will aid the line in obtaining strategical and tactical objectives.

The number of staff functions may vary depending on the need of the line and assistance required by the incident manager. Staff positions may have to be further subdivided into manageable units. Most agencies working an incident with up to 25 units can manage the incident with an ICS organization consisting of and Incident Command, Operations Section (up to five Divisions and Staging), Safety Officer, and a Resource Status Officer. There also may be a need for a medical unit, food and supply person, and depending on your area a public information officer (SEE FIGURE 3-2).

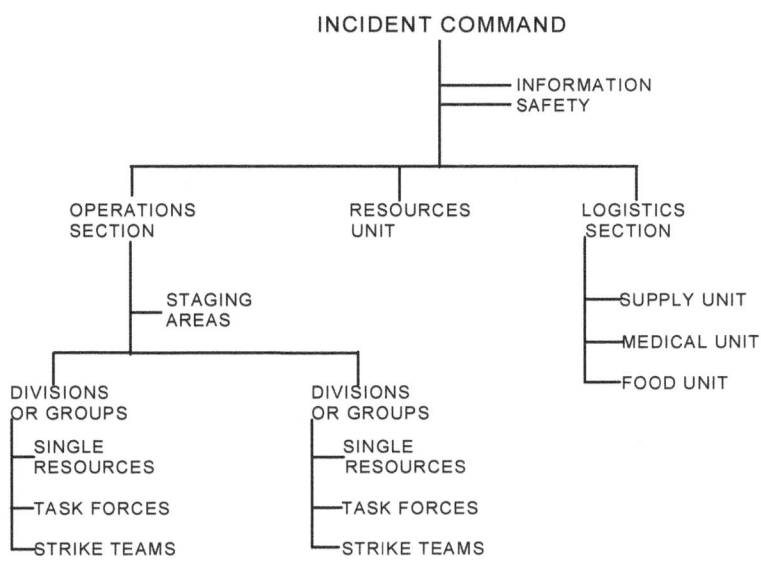

FIGURE 3-2

ICS has gone through many changes since its conception in the early 1970's to deal with wildland firefighting. During the next century ICS will most likely go through other adjustments. What will ICS look like in the future? That is a difficult question to answer because each innovation has brought with it some minor changes. However, the basic concept behind ICS and many of the positions have not changed since inception and we doubt that there will be any major changes in the near future.

One standard in all Incident Command System organizational versions to date is that no one person supervises more than five individuals, groups or functions at one time. This is the ICS "Rule of Five." This rule has been used in all versions of ICS since its development in the early 1970's under ICS-FIRESCOPE.

The Rule of Five has some exceptions in ICS. The functions of public information, safety, and liaison if grouped with operations, planning, logistics, and finance ends up with seven people reporting to the Incident Commander (IC). However, on most incidents you will not see all seven elements active at any one time. Yet, should an incident reach this point creating a command staff officer to manage safety, liaison, and public information can reduce the IC's span back to five (SEE FIGURE 3-3).

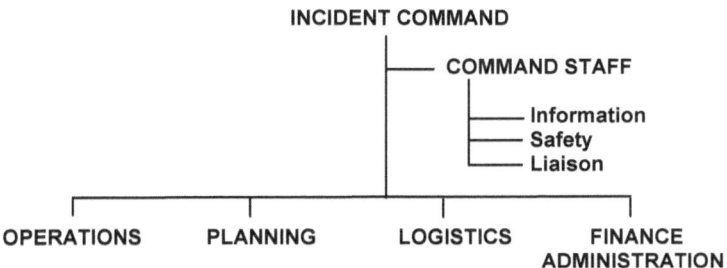

FIGURE 3-3

From a tactical standpoint, there are times when groups of less than five are needed to accomplish specific tasks. This is why the task force was created under ICS. A task force is a group of up to five different types of units or a group of less than five of the same type assembled to carry out a specific mission.

The Rule of Five was originally selected because of the span of control, ease of division, flexibility in use of resources, safety, and economy of scale. It has been, and remains, the best way to control a large number of resources and functions in an efficient and economical manner.

Practical ICS

ICS-type organizations are being used for virtually every type of incident that might be encountered by the emergency fields. The application of ICS depends largely on how people function in a given position. Therefore, in this chapter we shall not dwell on theory, but shall concentrate our efforts on how a person can function in a given position to achieve maximum effectiveness.

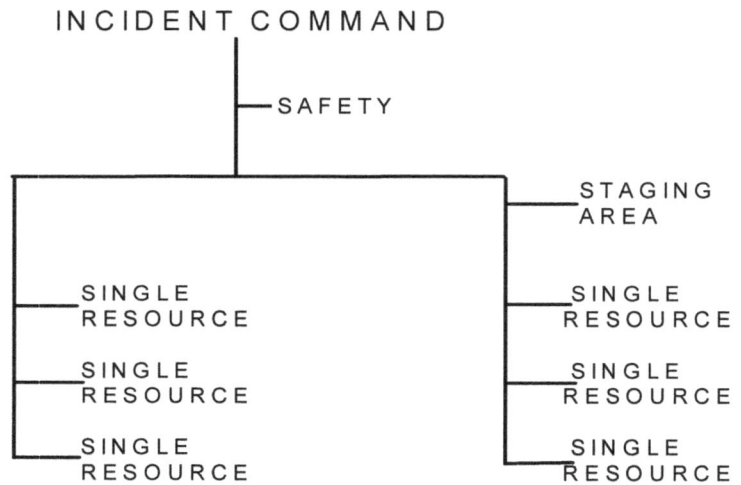

FIGURE 3-4

The key to a successful ICS organization can be found in one position - that of the Incident Commander. How the Incident Commander builds the organization may affect the outcome of the entire emergency operation. Traditional forms of emergency organizations are developed from an autocratic viewpoint, centered on the Incident Commander, then building the organization from the ground up. This is known as a centralized organizational use of ICS (See Example 3-4). The problem with building an organization from the ground up like our example is you can easily end up with a span of control exceeding five.

The centralized or ground-up approach can work very well so long as the incident and need for a larger organization do not develop too rapidly. Should this occur, there often is a considerable amount of un-necessary confusion because of the Incident Commander's span-of-control quickly exceeding the recommended five (SEE FIGURE 3-4). The way to avoid this is to build the organization much like many new businesses -- from the top down (SEE FIGURE 3-5). Select your executives first, or in our case, your key supervisors, to help manage and build the emergency ICS organization. The suggested approach for the Incident Commander to use is:

FIGURE 3-5

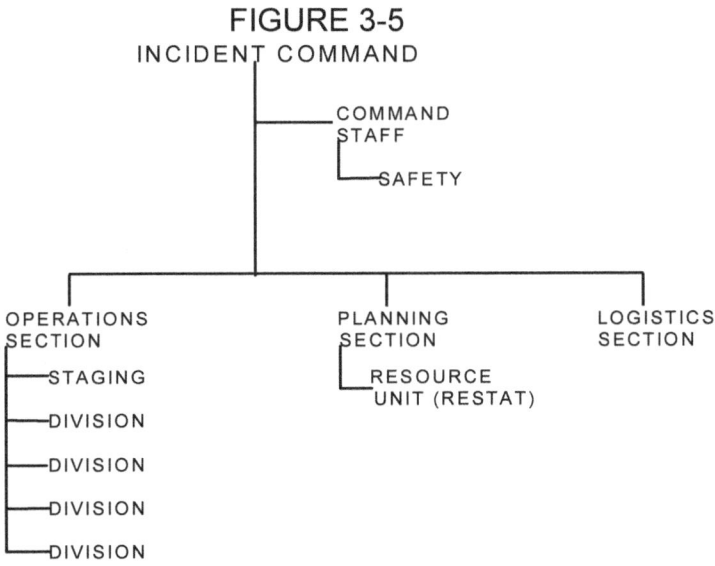

1. Establish your incident command post in an area conducive to the size and type of situation.

2. Assign individual unit supervisors to operational divisions and staging if necessary.
3. Appoint a Command Staff Officer, assigned to act as safety officer.
4. Appoint a Planning Officer, assigned to act as Restat, if necessary.
5. Establish other supervision officers as necessary, if the incident is to exceed three divisions.

   A. Operations Officer
   B. Logistics Officer
   C. Finance/Administrative Officer

Once you assign or appoint other primary supervising officers, allow them to establish additional organizational needs and manage those resources. This allows you, as the Incident Commander, to stay focused on providing the necessary support to meet the strategic objective.

Using this approach allows you to maximize the use of personnel toward accomplishing the strategical and tactical objectives, while minimizing the need to continually realign the organization and advise personnel of changes. This should also concentrate resources in the line part of the organization or action groups on the emergency.

Each time a need occurs, appoint an acting logistics officer to meet the need and manage the growth. For example, if you have the need of a medical group in the logistics section, appoint the Logistics Officer with the task of setting up the medical group. Then when the next logistical need arises give the Logistics Officer the task of managing the establishment of the next group. The same method can be used for planning and command staff functions.

An operations officer's position should be created only if the incident requires, or is projected to require, more than three divisions. The Incident Commander should, in most cases, be able to manage three divisions, one staging area, and a safety officer with a minimum of five supervisors or acting supervisors, excluding him/herself. The size of the agency and the number of supervisors will dictate who should be trained to fill what positions. It is advisable that all supervisors be taught to function as division or staging officers.

Remember that ICS is only a guideline; it does not come with any hard and fast rules. The concept is for the Incident Commander to build an organization based upon need. Building the organization from the top down accomplishes this primary organization responsibility with the least amount of confusion. On the other hand, building the organization from the ground up usually makes the Incident Commander's job more difficult and generally does not maximize the use of personnel. In the next few pages is a brief description of the more important ICS positions and how they can be of use to an Incident Commander.

Operations Section

The most important part of any ICS organization is the Operations Section. The Operations Section is the line or action group in an ICS organization. Without it, strategical and tactical objectives could not be achieved. The Incident Commander must give primary importance to see that sufficient personnel are allotted to Operations. This must be done even at the expense of other organizational functions, unless the functions are so critical to the support of Operations and without their implementation, the strategical and tactical objectives could not be achieved.

An Operations Section supervisor should be established when there is to be more than three divisions, two branches, or when staging, air operations, and line assignments are necessary. Normally on smaller emergencies, the Incident Commander handles this function.

Once an operations officer is appointed the individual is given the responsibility to achieve the strategic objective. When this occurs, the Incident Commander becomes more of a facilitator using the assistance of safety, planning, logistics, and finance, to anticipate needs and procure necessary resources.

Staging

Staging is an important part of the Operations Section. Generally, any emergency of any duration or one that calls for additional resources will require at least one staging area. The staging area should be established in an area that is large enough to accommodate the resources that are being requested. The location needs to be as close as possible to the divisions so as to provide needed equipment and personnel with the least amount of delays. On larger incidents, it may be necessary to establish more than one staging area in order to provide sufficient resources in a timely manner to the divisions or branches. In these situations staging can be coordinated by a staging area manager or, more preferably, the branch or division supervisor.

An Operations Section supervisor or Incident Commander functioning in that role may dictate how the staging areas are to operate. Some managers prefer to have control over staging, making all requests come through the Operations Section Supervisor/Incident Commander. This first approach keeps the person in charge of operations more aware of the total amount of equipment and personnel available. Other managers provide branch or division supervisors direct access to the resources in staging. The second method gives the staging manager more leeway in controlling the flow and requesting of additional resources. Either procedure can be effective depending on the type and size of the emergency, and the training and experience of the staging supervisor.

An effective staging supervisor will maintain a given number of resources in staging at all times and will establish reorder minimums. Reorder minimums are based on the estimated travel time from the time of request until arrival.

Depending on the situation, staging also may also establish a Rehab Unit for line personnel engaged in strenuous work. The Rehab Unit or Staging Officer may be given instructions to provide personnel rotations on a regular basis. This can prevent individuals from being overworked and improves the use of limited resources. A Rehab Unit should be a sub-unit of Staging unless the people in Rehab are not available for assignment. In this case, Rehab and the personnel assigned to it should be under the Logistics Section, Medical Unit.

Air Operations

To what extent air operations will play on an incident is difficult to determine. Generally, this depends on the size and type of incident and normal resources available to the responding agency. If air operations is to have a direct impact on the accomplishment of objectives, it is rightfully a part of Operations. If the use is only for observation and reporting progress of the incident, then perhaps the function belongs under the Planning Section, Situation Status. When observation and objectives' accomplishments are combined, the Operations Section should provide the coordination.

Branches

Branches become necessary when more than five divisions are needed. The complexities of Staging, Air Operations, and line functions make managing more than two divisions difficult. Branch functions can be diverse, such as triage, law enforcement, haz-mat, and firefighting, and when the emergency impacts more than one jurisdiction. The decision to form a branch is generally made by the Operations Section supervisor. Branch directors generally assume more command-type roles when they are appointed. Usually individual staging areas are required. When branches are implemented, it is usually because the situation has increased in size and scope. Branches often are not established early in an operation; therefore, division supervisors must be advised that branches have been activated, along with informing them to whom they are to report. Branches either should be titled by function or numbered.

During a multi-jurisdictional situation, the Incident Commander may decide that two or more branches are more advantageous than creating an Operations Section supervisor. The reason for this would be to provide each agency with more control over actions taken in their respective areas.

Branches or divisions also can serve to combine specialized equipment or personnel under one manager. This can reduce duplicate efforts in several different areas that can provide a more efficient use of these often limited resources.

Divisions/Sectors

One of the most important decisions when using ICS has to be at what point to establish divisions, sometimes termed sectors. During initial actions, the Incident Commander should designate someone from the first-arriving resources to act as a division supervisor. When more than three divisions become necessary, an Operations Section leader should appoint additional division managers as necessary. It is preferable that divisions be titled by letter designators.

Division supervisors are the persons who will supervise the tasks necessary to accomplish tactical objectives. They should be proven emergency managers who are aware of how to deal with normal incident problems and know how ICS operates. Briefings by the Incident Commander or Operations Section leader should include not only tactical objectives, but also strategic objectives and time frames, who they are to coordinate with, and where and how to obtain additional resources.

Division supervisors may manage single functions or a multitude of tactical objectives. Usually single function divisions are formed on smaller incidents. When an emergency covers a large geographical area, division supervisors generally are assigned a portion of that total area. They may have to coordinate several tactical objectives at the same time within this assigned area. The division supervisor is also responsible for making certain actions in the division are harmonious with the other divisions and support the strategic objective set by the Incident Commander. Preferably, a division supervisor should have a means of transportation. This may not be possible in the initial stages of an operation because of limited resources

A division leader may have up to 25 individuals or units assigned to in five groups of five each. When resources exceed these limits in one division, it is usually advisable to create an additional division. The use of pre-established strike teams or task forces can reduce some of the organizational and management problems with a division.

Strike Teams and Task Forces

A primary element of the ICS Operations Section can be the strike team or task force especially on larger or critical type incidents. Both a strike team and task force, derived from the military, are a group of resources assembled for a specific purpose or project. A strike team is a specific type and kind of resource, while a task force does not have to possess any commonality. It is a good idea to have inter-communication capabilities within a strike team or task force.

A strike team or task force should not be comprised of more than five nor less than three units with a leader, preferably with a separate own means of transportation. The idea is to maintain a reasonable span of control for the leader. Generally, a supervisor should be able to manage up to five units on most situations. However, numbers may be increased or decreased depending on the complexity of the incident, personnel experience and ability, physical separations, ability of the team leader, and demands (physically and mentally) of the situation from a resource-need standpoint.

The strike team or task force and its utilization concepts can be advantageous or it can create handicaps depending on the situation, resources available, time factors, and methods used to form the group.

Just what a strike team or task force in your area is comprised of or is used for is not as important as common knowledge shared by all agencies in your area. Be sure that time is taken before an incident to set acceptable methods for communication, standards for formation, leadership qualification, number of personnel assigned to individual units, etc. The ICS model strike team or task force will work and become an important part of the Operations Section provided it fits your area's needs.

Planning Section

Next to the Operations Section, Planning is the second most important element under an ICS organization. The Planning Section is the second most important element in an ICS organization because it provides much needed primary support for the Incident Commander and Operations Section, especially on a larger or more complex situation.

When the first need for an element under the Planning Section is established, the Incident Commander should establish a Planning Section manager. This person should handle any function needed under the heading of planning. If demands on planning begin to exceed the capabilities of this one person, the Planning Section manager should ask for or obtain additional qualified personnel to fill the necessary positions.

Resources Unit

The Resources Unit, termed RESTAT, is generally one of the most helpful elements to an Incident Commander in the early stages of an emergency. In this mode of operation, the primary duty of a Resources Unit leader is to keep the Incident Commander advised of additional resources that are enroute, in staging or assigned. In some cases, an Incident Commander may find it advantageous to have the unit leader not only track resources, but take over the acquisition of additional needs. Usually close coordination is required between this position and staging or the Operations Section manager.

Sometimes it may be necessary for RESTAT to have a network of assistants, called check-in recorders, to help track the arrival of additional resources.

Situation Unit

SITAT, as the Situation Unit is termed, is responsible for providing information on the status of the emergency. This includes providing the incident personnel with drawings or displays of the situation, and when necessary sending personnel to move about different operational areas to obtain firsthand information on actions of personnel and behavior of the involved elements. On wildland fires, the FIRESCOPE method of operation could be used to provide computer projections of potential fire spread within given time frames.

Documentation Unit

The Documentation Unit personnel serve as the file clerks and incident historian. Most agencies will find this an unnecessary position as an emergency assignment. The exception may be when the incident progresses to a point where federal or state monies may become available due to a State of Emergency declaration.

Demobilization Unit

Like the Documentation Unit most agencies will find this position unnecessary for most incidents. Yet, if an incident is extremely large and requires numerous resources a Demobilization unit is needed to assure orderly, safe, and efficient release of resources.

Technical Specialists

Typical ICS organizational textbooks list several specialists. The type and number of specialists should not be limited to what is found in any text, but should be dictated by the size, extent, and problems presented by the emergency. The knowledge of the technical specialist should be considered. When key individuals have as much knowledge as a specialist, oftentimes the appointment of a specialist is not conducive to effective use of personnel resources. Sometimes they would be better utilized as a part of the Operations Section. In some cases where they are under contract to provide specialized services, they may even be an operational division supported by public agency divisions and support personnel.

Logistics Section

The Logistics Section is what we term a secondary support unit on most ICS incidents. Unlike the Planning Section, which provides essential support services relating directly to line operations, the Logistic Section positions tend to be of a less urgent need. The larger the incident, the more resources are requested, and increased potential may call for more of a priority being given to logistic functions.

The build-up of this section should follow the same pattern as set for the Planning Section. When the first logistical need is required, appoint a logistics manager to fill the need and manage the growth of the section.

The Logistics Section is often shown on organizational charts as being divided into two branches: service and support. Most incidents will not require two branch directors; in fact, it may be possible for the Logistics Section manager to handle supervision of all functions because this is considered a non-emergency organizational element. The maximum number of logistical units is usually projected at six; therefore, based on non-emergency management principles of a controllable span being three to six, this would not be unreasonable. The appointment of one branch director to take charge of either the support or service units would reduce the span of control.

Communication Unit

On larger incidents the need may arise for a Communication Unit. The larger the emergency, the more important it becomes to have a Communication Unit. Radio systems alone cannot, and should not, be expected to facilitate good communications. It becomes the job of the Communications Unit to maximize the use of the radio, telephone system, and other forms of communication. On very large situations, it may be necessary to have a fully-staffed incident dispatch center set up in the field.

Medical Unit

The Medical Unit is not generally used for the treatment of civilian victims of the emergency, but is staffed to provide medical aid, transportation, and short-term care for injured and ill emergency personnel. If the medical personnel are involved with victims in operational areas they should be a function of the Operations Section. In some situations there may be a medical division in the Operation Section to deal with victim treatment.

Food Unit

One of the most overlooked items on an incident in general is often the feeding of personnel. To provide adequate support in this area, the Incident Commander or Logistics Section manager has to appoint a person to supervise this function long before the anticipated need.

Supply Unit

The Supply Unit orders and provides personnel for the Logistics Section, as well as tools and equipment for other sections. An important part of this section's activities is to provide documentation as to who receives supplies so the return can be assured. The Supply Unit also should handle any routine maintenance or repair of tools and equipment.

Facilities Unit

A Facilities Unit becomes necessary when an incident is going to require sleeping and sanitation facilities. Schools, campgrounds, churches and other similar buildings usually provide an excellent area for these types of support functions. Base activities have been modified by some agencies to be more of an active support function in the initial stages of an emergency.

Ground Support Unit

The Ground Support Unit is responsible for all equipment maintenance and repair, along with providing needed fuel. A Ground Support Unit should be established any time repair or service of apparatus and equipment is needed at the scene of an emergency. On larger incidents, the Ground Support Unit leader may work through the Supply Unit to obtain needed fuel and parts.

Some agencies may contract for specialized equipment. If that is the case in your agency, a contract coordinator assigned to the Ground Support Unit should become responsible for coordination of these resources, checking their condition, making sure insurance is provided, and keeping track of chargeable hours.

Command Staff

Most ICS organizations refer to Operations, Planning, Logistics, and Finance as "General Staff." General Staff, except for Finance, is responsible for carrying out missions that have direct impact on how strategical and tactical objectives are accomplished. Command staff personnel, on the other hand, assist the Incident Commander directly, by relieving some of the activities that are secondary and providing essential direct support.

A command staff officer is not often shown on ICS organization charts, although on some emergencies it may be necessary, because of personnel shortages, to have just one person appointed to command staff who handles all the functions.

Information Officer

An Information Officer can be of value even on small incidents if the media is on-scene. The need for the Incident Commander to keep track of the strategical and tactical objectives cannot be compromised. A trained Information Officer can remove a great deal of pressure that is often applied by the media. On larger incidents it may be necessary for the Information Officer to keep the press out of dangerous situations, or provide protective clothing along with guidance.

It is preferred that an Information Officer be trained in this specialized field; however, when the situation dictates any person may have to be pressed into this role. The less formal training a person has the more important it will become for the Incident Commander to approve the release of information.

Safety Officer

The role of the Safety Officer is important on most emergencies. One way to reduce injuries and improve safety is to have a person who specifically looks for hazardous situations. The Incident Commander cannot be expected to see and control all activities; therefore, injuries could be possible at any time. The Safety Officer has the specific responsibility to see that critical operations are closely monitored for safety hazards. This may require that the Safety Officer have any number of personnel assigned to assist in making the incident safe. In most ICS organizations the authority of a Safety Officer is second only to the Incident Commander.

One way to improve safety in the future is to have safety personnel investigate accidents and provide an objective report of the conditions and actions that led to the damage or injury. This information becomes extremely important for post-incident analysis.

Liaison Officer

The original intent of a Liaison Officer was to provide a person (agency representative) who could speak on behalf of a participating agency on all matters. However, in most ICS organizations today, the role has been expanded to include agencies such as building security, public works, telephone company, Red Cross, etc. Agency representatives have become so numerous on some incidents that they out-number the primary support personnel. This is one area where need should be closely examined before establishing a Liaison Officer or requesting an agency representative. Above all else, agency representatives should not be sent to an incident just because resources are available or in use. Generally, if there is a need to ask a question of any agency it can be handled by telephone or other means of communications by an appointed Liaison Officer. In most situations, personnel assigned as agency representatives can fill more important positions in the Operations, Planning or Logistics Sections.

Finance/Administrative Section

The Finance/Administrative Section is the least needed of all the sections on most emergencies. This is because most agencies have adequate methods of tracking chargeable hours for units on scene by use of internal records and/or recorded radio traffic with times provided either automatically or by voice insert. When an incident is beyond the capabilities of these methods then a Finance Section would be advisable. In addition, a Finance Section using ICS methods of recording time, claims and costs would be highly recommended when a state or federal State of Emergency is declared because reimbursement may be tied to good ICS record keeping.

PROBLEMS WITH ICS ORGANIZATIONS

Although we have attempted to demonstrate the effectiveness of ICS, there are some common inherent problems other than those we have covered that can and will occur during an emergency.

The most obvious problem occurs from lack of training and understanding of what functions are required of a given position. This can, to some degree, be overcome by operational briefings at the time of the emergency. But the most important aspect is that managers know the capabilities of their personnel.

Another problem that occurs with ICS is over-staffing of positions. Managers must keep a realistic view toward support functions. The major emphasis should be placed on providing sufficient line personnel and then filling staff positions as necessary. Oftentimes managers become so intent on filling staff functions because they appear on the chart that they severely drain the resources needed to keep the line part of the organization or action group effective.

One problem that is not often discussed in any textbook is attempting to make the incident fit the ICS organization. Sometimes sample organizations that are shown in ICS publications just make working with the system cumbersome. In situations such as these remember the organization structure is secondary to what works for the incident. Make the organization fit the incident, not the other way around. If you need a function that normally would be under planning or logistics in operations as the incident commander you have the authority to make it happen.

The last problem we shall mention in this chapter is the delegation of too much authority and responsibility. This problem occurs because of the pre-established ICS positions and training. Incident Commanders sometimes fail to recognize that they fill all the positions upon arrival at an incident, so to speak they wear all the hats. The individual in charge releases these positions or hats to other managers and supervisors only when the situation warrants such action. One of the most common errors is the creation of an Operations Section director on an incident where one is not necessary or at a point when the need is not yet determined. What happens is the Incident Commander effectively removes or shields the command position from the information intake and decision process. This affects the Incident Commander's ability to project the primary strategical and tactical objectives or meet long-range needs.

NOTES:

CHAPTER IV

CONTROLLING AND ACCOUNTABILITY

Many emergency managers are able to develop sound plans and build a suitable ICS organization to accomplish the strategical and tactical objectives with the time frames established. Both planning and organizing share the responsibility for bringing about the results envisioned by the Incident Commander. Plans are implemented throughout the organizational functions, but cannot ensure accomplishment by themselves. A manager must continually monitor the performance of personnel and the organization in relationship to the objective. This is achieved by enacting adequate controls along with maintaining accountability for personnel.

Every phase of an emergency situation is effected by controls. But despite their importance, there is a wide lack of understanding of what encompasses an effective emergency control as opposed to a poor one. Emergency controls differ from normal business-type monitoring that ensures quality and quantity to provide for adequate profits. An emergency control is designed or established by a manager in order to tell when a problem has developed soon enough to make corrections and still meet the strategic objectives within the given timeframes of the plan. In order to clearly identify some of the characteristics of effective emergency controls the following statements are provided and later explained.

> "Controls must not be too stringent, should be specifically required, need to focus on positive accomplishment, rest on measurability, should be kept simple, need to result in corrective action operate by comparison, need to be properly timed, should be based on seven factors."

Controls Must Not Be Too Stringent

One of the advantages of ICS is that personnel can be assigned to organizational positions that have pre-determined functions. This is especially true of support sections and units. However, in the Operations Section specific controls need to be en-acted that will ensure objective accomplishment without being so stringent that subordinates do not have sufficient flexibility to use their own initiative. Most ICS positions allow flexibility; in the Operations Section it may be necessary to go against this trend and have tight controls over critical objectives or activities. This is usually done even at the risk of being slightly annoying to subordinates in order to provide improved safety, better coordination or ensure attainment of strategic objectives. When stringent control of this type is needed, a manager should advise the subordinate of the reasons so that acceptance of the stringent control will follow.

Controls Should Be Specifically Required

An effective manager only enacts those controls that are needed to achieve the strategic objective of the plan. The simplest and safest approach to a problem is to let the person in charge of a specific objective or task, implement the necessary controls. A control informs the manager when a problem occurs which may cause the plan to be altered. It does not inform the manager of minor problems as long as personnel can enact needed adjustments within the scope of their assignment.

Controls Need To Focus On Positive Accomplishment

A control needs to have a positive aim. Control should ensure the "right" things happen and must not be negative or keep things from going wrong. The manager who realizes that awareness, not

repression, is the most important purpose of controlling has gained valuable knowledge.

Negative's in controls shows lack of faith in
subordinates and should not be treated lightly.
Negative controls hide the objectives; they are
psychologically self-defeating and do not provide
direction.

Positive controls tell a manager the objectives are
being met and good things are occurring. A good
control would enter at mid-point in an operation. For
example, a good control would say, "Let me know
when you reach this point;" a negative control for the
same operation would say, "Let me know when you
have a problem." The point is that the negative control
may lead to the assumption that everything is going
great until it is too late to make any adjustments. A
negative control seldom prevents bad things from
occurring and does not cause good things happen.

Controls Rest On Measurability

Controls measure progress toward strategical and
tactical objectives. Therefore, an effective control has
a quantitative character to it. A manager should use
controls to inform supervisors when progress is not
being made or when corrections are necessary in
order to achieve the objectives. The effectiveness of
controls is lost or diminished when they do not have
the ability to measure accomplishment or are not
based on quantitative measurements such as time.

Accomplishment-oriented controls are the ones where a subordinate or supervisor is required to check back when a given amount of work is completed. Depending on the length of the task or objectives, any number of controls could be established. Usually it is not advisable to provide more than one control at a time. When an employee reports completion of one segment at a control point, the manager should proceed to establish a second control and so on.

Another method of measuring progress is by the use of time. If an operation is extremely lengthy in nature, it may be much more advisable to use time as a control. This type of control requires a subordinate to report back at pre-established timed intervals. In most cases the manager best initiates this type of control. Managers should make notes as to when to check progress on timed intervals. This means the manager must use a watch as a control tool.

Controls Should Be Kept Simple

Controls, as with many other emergency activities, should be kept simple. Simplicity is the key to success. Managers should avoid what is termed a double-decker control. Double-decker controls require an employee to report back unless this or that has occurred, then to take this action. This leads to confusion on the part of both the manager and the employee. The employee is uncertain what action to take and the manager has a problem remembering or tracking what instruction was given to whom. The simplest control is based on mid-point or timed reports and should be used whenever possible.

Controls Need To Result In Corrective Action

Controls should be designed not only to inform a manager of when a problem is occurring, but also to cause change. The ideal control results in corrective action either in the organization or the elements of the plan. Although a control should not operate in this manner often, when it does a change of some kind should result. The outright sign of a defective control is that nothing happens, when a problem is endangering accomplishment of the strategical or tactical objectives.

Controls Operate By Comparison

Effective and efficient controls require the adoption of objective, accurate and suitable standards of measurement. This is not always possible with certain emergency operations. But when there are set procedures, such as with an ICS organization, it becomes simple to compare what is taking place with the position standards that are generally provided. Thus managers should give attention to the ICS position definition and detection of the exceptions. Only large variances in ICS that affect the general efficiency of the team approach should be controlled.

Controls Need To Be Properly Timed

Times relate to creating feedback at the proper time and place. Controls operate best when they are inserted: where something is likely to happen, where change of some kind should occur, or when critical performance is required.

Controls should be timed so that they do not place a strain on the ICS organization. The action of controls and the building of the organization must be compatible.

Controls Should Be Based On Seven Factors

The number of controls that may be needed will vary depending on the situation and seven impact factors. These factors are as follows:

1. The employee's training (experience and ability)
2. Complexity of the work
3. Length of the job being performed
4. Difficulty of the duties
5. Number of other personnel involved
6. Physical separations
7. Your ability

Employee's Training

Where training is concerned, primarily we are speaking of experience and ability. The more experienced a subordinate, the less control a manager has to have. However, if there is a specific procedure out of the norm, it may be extremely unwise to be lax in the use of controls. When ability is considered, we must know something about how the person being supervised reacts to certain types of situation in order to control the activity without becoming aggravating to the employee.

Training personnel in what a control does for a manager can also be extremely helpful in obtaining adequate feedback when it is necessary. In turn, this can decrease the amount of information we must sort through on a very complex problem.

Complexity of the Work

From any standpoint, there are not many jobs that are more complex than that of emergency management. Even a very routine or simple problem has many different facets that must be at least evaluated, if not acted upon. As an emergency escalates, the management problems become more complex and likewise the difficulty of enacting controls. A manager in these situations may find that it would be an advantage to reduce the number of subordinates, thus limiting the number of direct subordinates and permitting more effective use of controls.

Length of the Job Being Performed

The longer an operation takes, the more opportunity to enact controls; but also the more likely the job may be forgotten until it is recognized as a critical problem.

Managers, not subordinates, must use a positive tracking method to account for controls and the checks and balances they provide.

Difficulty of the Duties

Many of the jobs on an emergency scene are extremely difficulty due to the amount of physical and mental stress. Managers must take into account how much difficulty is involved with the duties being performed and enact controls that will not add to the burden of the tasks.

Number of other Personnel Involved

The larger an organization becomes the more difficult it is to control activities and obtain information in a timely manner. When more than one subordinate is required to perform required duties, one must be in charge of the operation and is responsible to report back to the manager in charge.

If Possible, the span of control must not be exceeded on emergencies. ICS is built on the span not exceeding five with three being ideal. Managers should be aware that fragmenting resources could lead to the span of control being exceeded. On larger situations where more personnel or equipment is in use, it may be advisable to require subordinate managers to report on the status of the number of people reporting directly to them. This is one way of controlling the span of control in the organization.

Physical Separations

When a manager is working where there is direct contact with subordinates, the job of controlling activities is much easier. That is why, in some office-type environments, the span of control is seven or more. However, emergency management usually involves physical separations, i.e. interior from exterior, front to rear of structures, block to block, ridge to valley, etc. Whether these separations are only a few hundred feet apart or miles, the problem of controlling activities is compounded by the fact that the manager in charge may not be able to see any progress. Thus, more communication becomes necessary and in turn, the number of people or activities one can effectively control must be reduced.

Your Ability

The final factor that affects control personnel or activities is your ability. What might be too many for one person to control in a given situation may be simple for another. Your ability in this area has a direct correlation to your training and experience as an emergency supervisor or manager. Generally the more opportunity you have to exercise control skills, the more proficient you are likely to become. Part of control experience is recognizing that a problem is increasing and complexities are compounding, which in turn calls for increasing the organization but reducing your span of control. Most experienced emergency managers agree that the ideal span of control is three with five being a maximum.

CONTROL METHODS

The factors, which impact controls and their designed use or application, should have been clarified. Control methods are the types of procedures used by a manager or supervisor to monitor progress or the lack thereof. Unlike typical management situations, on emergencies only three basic types of controls have been found to be effective. They are...

1. Visual
2. Periodic reporting
3. Inspection

Visual Controls

The most effective of all emergency control method is visual. A visual method of control is used when you can directly observe the activities that are being implemented and their progression towards completion. This type of control is only used when operations are within line of sight.

A visual control is very easy to utilize on a simple incident where a limited number of resources are committed and where more activities are centered within a small geographical area with few, if any physical barriers. As a problem becomes more complex or spread out, your ability to constantly observe operations is split respectively. There are some problems that can occur because of this division of attention.

One is tunnel vision. Should one area develop a problem, as a manager you may tend to focus all your attention in this area, especially if you are in close proximity to begin with. To avoid this common problem managers must learn to use distance in order to widen a field of vision. A good rule of thumb is to locate your command position at least one hundred feet from the work being performed.

Another problem that can occur is over-controlling. Over-controlling happens when managers or supervisors release little, if any, authority and responsibility to subordinates. When visual controls are used the same release of authority and responsibility must occur; otherwise, if the problem expands the manager will tend to become overwhelmed. In addition, without this release of authority and responsibility, employees are not permitted to develop their management skills.

Sometimes the problem of "oversight" can develop when visual controls are used. Oversight occurs when a manager looks at an activity and tends to see only what they want to see, oftentimes-overlooking glaring errors, inadequate personnel, shortages of equipment, etc. Again, as with tunnel vision, distance can generally solve this type of problem.

Periodic Reporting

The most over used and abused type of control method is that of periodic reporting. One of the primary reasons for this is that managers or supervisors fail to convey strategical and tactical objectives as a part of the plan. Another reason is poor communications practices within the organization and not teaching supervisors how to use periodic reporting controls. The best way to carry out a reporting type control in any situation is to have an individual describe activities in a face-to-face progress report. During emergencies, managers should not hesitate to recall supervisors for a periodic report especially when radio communications tend to confuse the issues.

When periodic reporting by radio is the only means of controlling activities because of physical separation, communications must be kept simple and to the point. Excessive non-essential communication has to be eliminated. As a manager, a person should attempt to control the operation by using known points of progression in the task process. Ask specific questions regarding this progression. For example: Has this been completed, how long before this is done, etc. Avoid questions such as "how are you doing" or others that might lead to unnecessary communication on the misconception that the operation is progressing as planned.

Inspection

Often on emergencies an inspection of an area or operations may be the only method that will adequately provide the needed control. The best way to accomplish this is to have an aide or runner who becomes your roving eyes during the emergency. This person can be sent to check on conditions in the different areas and provide a firsthand report of progress.

In certain cases, due to conflicting information being received from subordinates or problems developing beyond one's ability to comprehend, an actual inspection by the manager in charge may be of a necessity. In most cases, it is not advisable to completely abandon a command position. Instead, someone should be left at this position and the manager should quickly survey of the situation. Care must be taken not to become involved with activities. A survey by air is most effective, but not always practical or possible. When making a ground survey learn to look, make drawings, take notes and say nothing to anyone. In this manner you increase objectivity and reduce tunnel vision and oversight.

THE RELATIONSHIP OF CONTROL METHODS TO EACH OTHER

The larger, more spread out and complex the incident the more difficult it becomes for a manager to control the organization and the established objectives. An effective manager learns very quickly that most emergencies require a blending of all three-control methods to have truly effective controls. Each of the control methods does have stand-alone qualities, but a single control method would be limited to only a small number of incidents. Emergencies of increased size and scope cause the Incident Commander and other managers to become more isolated from the action group, causing a blending to control methods in order to focus conclusively on the strategical objectives.

Focusing on strategical objectives places more importance on periodic reporting by subordinate managers. Thus subordinate managers have to become the tools for visual and inspection methods of control. Delegation of authority and responsibility for controlling activities must follow with the development of the plan and the organization.

Controls on smaller situations, where the Incident Commander is also the Operations Section leader, operate at only the tactical level to ensure obtaining the strategical objective. Controls on larger emergencies operate at two levels, the strategical (Incident Command) level and the tactical (Operations Section) level. Without this two-tier approach, controls will fail to function as they should, possibly leading to errors, poor coordination, and failure to obtain strategical and tactical objectives.

ACCOUNTABILITY

Accountability and the tracking of personnel assigned to emergencies are becoming increasingly important during emergencies. Agencies across the country are finding themselves pressured to improve occupational safety at emergencies.

We believe that every emergency agency command system should have some method of addressing accountability for personnel. Still, the system used must not be so cumbersome that command officers tend to focus on personnel accountability alone. There are three ways to address accountability for personnel. One method is accounting for the location of every person, known as individual accountability; another is to account for unit assignments and the number of people, termed unit accountability; and the final method is a combination of both.

Individual accountability is the most effective means of obtaining the location and name of every individual on an emergency scene. Yet, these types of systems tend to be cumbersome and often cause command officers to be consumed with individual names and tracking their locations. This often causes more important strategical and tactical objectives to become secondary considerations. Oddly enough, this in turn creates a safety problem that requires individual accountability. These systems then become self-fulfilling prophecies -- the more you use them, the more you need them,

Some people would have us believe that individual accountability is a new idea. Actually, I remember one such method outlined in a publication printed in the 1940's (I can't remember the title). More recently, in the 1970's a system of this nature was used for a short period under ICS FIRESCOPE adaptations for building fires.

The passport system is the most recent individual accountability system. The passport system is nothing more than a rebirth of individual accountability. This type of system, rather than requiring recording of names or name cards handed to an officer in charge of an operational area, passes magnetic or Velcro name tags from officer to officer. This method of accountability works well if the names are held at the check-in point on an incident. It becomes difficult to work with, and distracts officers, when individual names have to move from officer to officer as assignments change.

To be effective for an Incident Commander, a passport type system requires at least one person to staff the resource status unit (RESTAT). Should other subordinate officers, such as division supervisors use the system an aide should be assigned to each area manager just to track the people. For most departments the inherent problem with this system is the number of people it takes to staff the extra positions. Used at the top of the emergency organization with a resource status officer or unit the passport system can be effective.

ICS is built on the principle of unit accountability. The reason for this is that units or companies are much easier to track than individuals. There is little argument with this statement. The difficulty is identifying how many people are part of a unit or company. This problem can be easily solved by training Incident Commanders and subordinate officers to simply record the unit or company number with the number of people assigned; i.e., Engine 2 with 4, Division "A"; 2nd floor, Eng. 5 - 3.

Certain individuals will argue that when tracking the number of people we need to know specifically who is where and doing what. From an incident management and safety perspective this is not necessary. In particular "who" is not important, the "what" in generalities may be. If you will, try to relate firefighting to fighting a battle in a war. During a battle the officer in charge of the incident or area needs to know personnel numbers and what companies are advancing. Should a problem occur that might involve injuries or death, the officer needs to know what has happened and how many are down and where. Also, can the strategical and tactical objectives still be met? If not, will this have an impact on the safety of the rest of the operation? The idea is to get sufficient personnel to the problem area quickly. Then assist in the assignment or rescue, care for injuries, and finally, account for "who" is down or in trouble.

Perhaps the best method of accountability is a combination of the unit and individual systems. Many

ICS agencies have used variations of a combination system for years with great success. The method used is simple and inexpensive. Upon reporting, the officer of each company provides a 3" X 5" card with the unit number to the Incident Commander, Staging Officer, Resource Unit, etc. The card also lists the names and ranks of all individuals assigned to the company. This card is held at the check-in point. Officers using the unit/company number with total personnel count; i.e., Engine 4 with 2 then track companies assigned to operational areas. If a problem does occur the quick check is by unit or company number and number of people. The back-up or secondary check becomes the card whereby every person is accounted for by name.

A combination accountability system is very easy to carry out in an all-career department. In volunteer or combination-career-and-volunteer departments, you may have to alter some operational procedures or develop alternate check-in systems to develop a combination accountability system workable. The simplest way to ensure such a system work is to have all personnel report to the scene in a agency unit. This makes tracking simple and all-inclusive. If your department allows personnel to respond directly to the scene in a private vehicle, then a clipboard sign-in sheet may be useful. Once personnel sign in on the clipboard the work groups should be established on the clipboard as a permanent record of assignments; i.e., Jones/Smith/Doe - Eng. 2.

ICS provides us with one other element that helps with accountability and that is the Rule of Five. No more than five units or individuals should be directly subordinate to any one individual. Although there may be some cases where more than five people report to one person, such as hand crews, the basic emergency organization should be built on the Rule of Five. I know of no organization that uses ICS and adheres to the Rule of Five that has ever had a problem with accountability; this includes situations where numerous firefighters have been inside a building during a catastrophic building failure.

It is important that we maintain accountability for personnel. Yet we should not establish systems that distract from dealing with the management of the emergency. When any system used during an emergency can distract officers, it has the potential to do more harm than good.

When tracking of individuals becomes more important than tracking results, and the safety as well as efficient use of resources, something is wrong. Tracking systems that are designed to protect individuals must not have the potential to contribute to their injury. ICS has proven personnel accountability system using the Rule of Five. This system is simple, inexpensive, and most importantly, does not distract officers. We suggest you look at the past and present history of on-scene accountability systems. We believe you will find ICS has a practical solution. The first step is to adopt ICS as your department's Incident Command System.

NOTES:

NOTES:

CHAPTER V

COMMUNICATIONS

Communication is the key to making the plan effective, the organization functional, and the controls operational. When building any emergency system the purpose is to take these inter-dependent parts and form a management structure that is designed to deal with the situation. A plan becomes the foundation, while the organization is the framework, and controls serve as the roof; but what s this a system is the communication. By themselves the first three components are not designed to stand alone, but with communications each can operate independent of one another. Communications is the element that integrates our diverse but inter-related and inter-dependent parts to form our management system.

Communication is the method by which the plan is conveyed, the organization structure identified, and the controls enacted. A manager can plan, organize, and design controls with poor or ineffective results if they can not adequately communicate. Communications is a word that means different things to a wide cross section of the emergency community. Webster defines communication as: "the act of transmitting; a giving or exchanging of information, signals, or messages by talk, gestures, writing, etc." But to most emergency personnel, it is the transmitting of information by use of the radio.

To the emergency manager communications has to be much more than the radio. What transpires between the Incident Commander, supervisors, and subordinates through communications has to be clearly defined, sufficiently discussed and acted upon decisively if performance is to measure up to expectations. Emergency communications is performance oriented. The power at the command of managers, supervisors, and subordinates cannot be released, directed, or coordinated without a network that can, and does, communicate vital and individually meaningful information.

Communications has been called a science, an art or a skill. It does not matter to which vehicle you feel communications belongs, so long as you recognize that emergency communications is a learned medium and what we do from a communications standpoint will have a major impact on all incidents.

How many times have we in the emergency field heard the words, "It was a failure to communicate" or "a communications breakdown" as the underlying reason for an error? It seems that each time we objectively conduct a post-incident analysis, communications or the failure thereof can be traced to well over 50% of the problems that were encountered as the operation progressed. The reason for this is that the ability to communicate is too often taken for granted, especially during the first minutes of commitment.

There are approximately 600,000 words in the English language. The average educated adult uses about 2,000 words in daily usage and the 500 most frequently used have 14,000 meanings. This emphasizes out an important thought to remember. Words do not have meanings; how people interpret and translate them into action does. The radio, although an excellent tool, is more often than not abused or misused causing problems.

Communication is the key to making any element of the management system operate smoothly. Communications can be broken down into three basic areas: theory, functional concepts, and daily operational procedures.

THEORY

Agencies should address and adopt as much of the theory in regards to ICS as possible, the most important of which is common terminology. Common terminology is essential for any management system, and especially one that could be used in joint operations by numerous diverse users. Agencies should adopt as much of ICS terminology and corresponding definitions as possible without making modifications.

Another ICS theory is that separate radio nets are available with command vehicles having the ability to scan secondary channels while constantly monitoring and transmitting on primary frequencies.

All communications between organizational elements at the incident should be in plain English. No codes should be used. This is done so that there will be no confusion between differing agencies. Eliminating codes will improve almost any communication network; but the feasibility of numerous radio nets may not be practical for agencies that can neither afford such equipment nor have not developed sound functional concepts and/or operational procedures.

FUNCTIONAL CONCEPTS

To produce communications that are practical in many agencies there may have to be a major departure from present informational methods.

Functional communications must be considered primary to command. This means that "unnecessary talk" needs to be substantially reduced or eliminated and communications confined to only essential messages. The exception would be an agency with a car-to-car or low-priority tactical frequency.

A discipline, in agencies with one or two radio nets, where it is common for everyone to be competing for air time will complicate any communication network and destroy the concept of being effective with ICS communications, especially during a multi-agency, major emergency.

Functional jurisdictions must teach personnel that only primary managers in key ICS organization positions are to be using the radio and they need to use the same principles during daily operations.

OPERATIONAL PROCEDURES

What is done as a routine will have a significant impact on the use of ICS communications. Operational communications at an incident should be managed through the use of a common communications plan and, if necessary, an incident-based communications center when an incident is extremely large. All communications between organizational elements should be by pre-established radio nets, on-site (cell, private or hard-wire) telephones, public address and/or public telephones. The use of microwave radio and cellular telephone communications should not be overlooked.

Operationally, it is preferable that agencies have available five basic radio frequencies for use on large incidents, one net each for command, operations, support, ground-to-air, and air-to-air. The command net links together the Incident Command, Operations, Logistics and Planning managers, and other key staff members.

The operations net is often termed a tactical net. The purpose of the operational net is to tie the different branches, divisions, and units in the Operations Section together. There may be a need for several different tactical nets if the incident is divided into diverse geographical areas.

The support net is generally established to handle status-changes for resources. This would include requests for additional personnel and equipment, the assignment of these resources to support or operational areas, and relief, if needed.

The ground-to-air net is a tactical frequency used to coordinate ground-to-air traffic. Air-to-air nets are used to coordinate the different air resources and their activities.

Using routine operational procedures, functional concepts, and combining them with theory, ICS communication scan, and, will vastly improve your ability to communicate on any emergency no matter what type or size.

Problems in Communication

Problems in communications arise because we forget that our individual experiences are never the same as the people with whom we work. Managers with more in-depth knowledge will inadvertently omit key elements in an order or directive, leading to misunderstandings. This occurs because usually the wrong mode or method of communication is chosen. Remember no two people ever hear, or for that matter see, things in exactly the same way.

Perception or what we see is often mixed with oral communication from differing locations on an emergency causing breakdowns. Problems or situations viewed on opposite sides for example can produce remarkably similar circumstances with little or no interconnection.

People communicate in a variety of ways. They do this by physical touch, by visible movements of portions of their bodies, and by symbols either spoken or visible. Failure in communications is often due to a misunderstanding of symbols or inadequate transfer of perception of these symbols. During emergencies the loss of any one of the following elements of communication is most likely to lead to communications breakdown: Spoken symbols, visible symbols or visible body movement. Note: If two of the three elements are visible, it is reasonable to expect that if you or your agency relies heavily on the radio that sooner or later a communications breakdown will occur.

Spoken Symbols

Spoken symbols have the most impact of all the communication elements. More misunderstandings occur because of what is said than found with any other element of communication. Spoken symbols during emergencies are those very important words that clarify and/or release authority and responsibility to another person. The simple stating of things like, "you are in change," "coordinate with," "report to," etc. can have a dramatic impact on the outcome of emergency operations.

Spoken symbols on emergencies can be said to erase doubt in the minds of subordinates. Oftentimes these symbols are action verbs that explain what has to be accomplished.

Another use of spoken symbols is found when the radio is the only means of communicating. A spoken symbol in this case attempts to tie individuals together by referring to common knowledge shared by both. Examples are; past experiences that could be applied to the present situation, training methods, step-by-step directions, etc.

How we say something and the tone of our voice has an impact on personnel. Excitement on the part of an Incident Commander while communicating can cause violations in even basic safety precautions.

Visible Symbols

Visible symbols are simply what the manager and subordinate can both see. Maps or drawings used to clarify communications are excellent examples. The old saying "a picture is worth a thousand words" should not be overlooked when managing emergencies.

Visible Body Movement

Visible body movement is often referred to as non-verbal communication. Visible body movement is most often thought of as "pointing" to identify places of common reference. But, non-verbal communication is also the look in a person's eyes, facial expressions and body position, all of which can effect the meaning of our words or the actions of personnel.

Watch people when you communicate. If their eyes are glassy or very wide you may not be communicating. Are they twitchy? They may be uncomfortable with what you are saying. Can you see a look of confusion in their facial expression or is it disagreement? These are but a few examples; there are others so watch for them to become a successful emergency communicator.

Timing and Space and Their Communication Impact

Timing and space will effect communication. A person must feel comfortable in order for communication to be effective. Timing is important. When possible, avoid communicating with resources that are not on scene. At the very least, they should be staged nearby. This will cut down on the need to change orders or correct an action that is not based on the current situation. Know when to communicate important information. Provide only what is necessary at the appropriate time.

Space has a large impact. The closer you are to a subordinate the easier it is to read the person or provide the correct type of communication. For example, during an emergency personnel can become very excited or apprehensive; a hand on the shoulder can have a calming effect. This is a type of communication that is not talked about in most materials on emergency communication; the impact though can be dramatic.

COMMUNICATION MODES ON EMERGENCIES

Generally it is the mode of delivery used by the Incident Commander that sets the stage for communications during an incident. The mode relates to the manner in which a manager or supervisor communicates with other personnel. Modes are identified as: directives, assignments and combinations of both.

Directives

A directive is very specific. It answers the questions of who, what, when, where, why and how. Who will perform the task, what will be done, when it will be accomplished, where it will be done, the reason for the job, and the specific procedures to use.

The advantage of directives should be obvious; they allow very little room for error or deviation. Directives work well for managers when there are limited resources and minor problems to be dealt with in a small geographical area.

However the disadvantages on emergencies usually far exceed the advantages. Besides not allowing individuals to use any initiative and not releasing any of a managers authority and responsibility, the following drawbacks can occur:

1. Too much detail is required
2. Keeping track of activities will be difficult
3. Numerous changes may be required
4. Feedback may be excessive
5. Coordination may suffer
6. Incomplete or insufficient information will be provided
7. Control may be lost, as well as an effective span exceeded
8. Reorganization may become necessary

These items say nothing of the dependency that subordinates may form upon guidance from a manager.

Assignments

An assignment is much broader in scope; and unlike a directive, it allows personnel to take individual initiative. But even more importantly, an assignment permits the manager to take full advantage of the knowledge that others may possess. Assignments require that the plan be conveyed, the organization identified, and controls established. They establish areas of responsibility and authority. This is the largest difference between a directive and an assignment. A subordinate working under an assignment can take whatever steps are deemed necessary to accomplish the objectives, relieving the manager to concentrate on the overall operation.

The advantages of giving assignments are many, the most important of which is the ability for subordinates to take action based on the plan but without specific direction. This permits personnel to take a course of action that will accomplish their duties in the safest, easiest and least time-consuming manner. The more spread out the incident becomes, the greater the advantage will be gained by this type of communication. The other advantages are;

1. Improved coordination
2. Controls are easier to maintain
3. Objectives are kept in sight
4. Excessive changes are not usually required
5. Feedback is kept at a minimum
6. The command capabilities of all personnel are improved

The only disadvantage, if it could be termed that, to assignments is the loss of knowledge as to what specific actions that are being taken. But then, a good managers will make fewer decisions than their subordinates.

Combination Communication Modes

Not all communications fall into the assignment or directive mode, sometimes there is a need for a combination of both. The combination mode is used when a specific is required of a person carrying out an assignment. Authority and responsibility is still released, but with one directive-type action required.

The reason for a combination mode is as follows:

1. The Incident Commander has knowledge of the situation beyond that of the subordinate
2. Coordination requires a specific action
3. The Incident Commander does not know the capabilities of the subordinate
4. Another action could endanger personnel beyond reasonable limits

The more knowledgeable a manager or supervisor is of the capabilities of subordinate personnel, the less likely a combination communication mode will be necessary.

Common Modes Problems

The most common error in giving either assignments or directives is omitting important information. Personnel undertaking any operation cannot be expected to complete the necessary functions correctly without adequate information.

Oftentimes managers create confusion on the part of subordinates without even realizing they have done so. An example is when a directive is given followed by "you're in charge" or "do what you think is necessary." This tends to confuse the situation by making the subordinate wonder what is supposed to be done or accomplished. The same thing is true of assignments when no objectives, time frames or controls are communicated. In either case subordinates should learn to obtain clarification before proceeding with an unclear communication. This can be accomplished by repeating the communication back to the manager, or identifying the objective, time frame and controls as you see them and conveying them to the manager.

METHODS OF COMMUNICATION ON EMERGENCIES

The problem of communication on emergencies comes down to two basic factors, clarity and conciseness. If a communication is clear, it is put together in a logical manner progressing from point to point. If it is concise, then it delivers the message of what is to be accomplished. Communication on emergencies can be handled through anyone or combination of five methods: radio, phones, runners, written messages, or face-to-face communications.

Radios As A Means Of Communication

The radio is the most common, easily accessible, and convenient method of communication. It is also the most abused, overworked, and disastrous tool ever invented when utilized incorrectly by the emergency manager. If these two statements appear to be in conflict they are, just as the radio often is during an emergency. When analyzing past situations, the most prevalent communication problem is the use of the radio.

Discipline is the answer to many of our radio communication problems along with the recognition of when not to use the radio. Discipline involves what is known as communications protocol or policy. Every agency, no matter how large or small, should have some written policies on what priorities are to be placed on certain radio traffic. Generally, it should be policy that use of the radio is primarily designed to aid key supervisors in managing an emergency situation. Other personnel should not routinely utilize the radio unless contacted by a command manager.

Supervisors need to recognize when to use the radio and when other forms of communication would be more effective. Generally those in charge should be aware that detailed communications, such as those in a directive, are not suited for the radio. Likewise, assignments to key personnel need to be made on a face-to-face basis. The reasons for this should be apparent. The radio, although efficient, is not as effective as other methods because it does not allow a free exchange of information from one person to another. The exception to this would be very simple communications involving one task.

The radio is an excellent tool for a manager to receive feedback and control activities; but as a sole means of managing communications on an incident, it Is less than adequate. A major area of agreement concerning the radio is that it saves time. However, this also is a misnomer, when supervisors consider the number of errors that are made when only the radio is used to communicate. Errors from strict use of the radio can cause longer delays, which in turn create confusion and more involvement from a time standpoint.

Phone Systems as a Means of Communication

The best means of transmitted communication is by a phone system. There are three basic types; public, cellular, and portable.

Public Phones

One of the best and most often over-looked communication resource is the public telephone system. Public phone systems classification includes any equipment that may be used to call numbers outside one area. This could include private phones owned exclusively by a private party as well as a public utility system. The only limiting factor in the use of these systems appears to be training and educating of emergency personnel as to their operation and advantages.

Training on different types of systems creates one of the largest problems. Home or single-line installations are simple, but business systems vary. Most business phone systems require some type of code or prefix before being able to dial outside numbers. This usually is a single number such as eight or nine but can be a coded number of three or more digits. Some systems require one procedure during business hours and a different one after hours.

The use of the phone in command situations permits a much freer flow of information between the Incident Commander and the communications center than that can transpire over the radio.

Caution should be urged and control exercised in regard to who can contact the communications center by phone during an incident. If some type of control is not exercised the communication center can become paralyzed. This is especially true on large incidents where they may have to use the phone to make outgoing calls to order resources. For this reason every communication center should have at least one line that is for outgoing calls only. Special lines of this nature can be ordered from the local phone company.

There are ways to get more mileage out of the phone system. Many closed systems have what is known as conference dialing where one or more parties may be on the line at the same time. Likewise, public systems often have what is known as a conference operator. This person can put you in contact with many parties at one time. The only criteria is that you must know all the phone numbers you wish to contact and how to contact the conference operator. Normally the conference operator can be reached by calling the phone system operator and asking to establish a conference call.

Cellular Phones

Cellular phones can go be even a more useful tool. Most agency staff vehicles are now equipped with cellular phones that can pick up transmitting sites from almost any location. The major limitation of cellular phones other than lack of cell-service in some rural areas is their ability to cause gridlock in the phone system. On major emergencies so many people sometimes use the cellular phone that communication centers are often overwhelmed.

Satellite Phones

A satellite phone can prove to be invaluable in areas without cellular service or where such service is intermittent. The largest drawback to these phones is cost of both the phone and air time. Yet, in areas without other phone services this may prove to be a small obstacle.

Portable Phone Systems

To solve the problems of radio frequencies becoming over-loaded with excessive traffic, some agencies have designed and constructed or purchased self-contained portable phone systems. These systems are battery operated, battery assisted, or sound power. Their use can be extremely helpful in running an incident where large support functions are set up in one location. All the systems operate in much the same manner. The only difference is the maintenance factor required on battery or battery-assisted systems. The most preferred system is the sound power type.

Runner as a Means of Communicating

Runners or aids have their advantages and oftentimes are more effective than the radio or phone. The way to maximize this resource is to avoid long, complex orders and to allow supervisors feedback through the runner.

Long and complex orders are apt to be misconveyed when using a runner. Anyone who has played communications games will attest to the fact that something is invariably lost in any relay of information. The more complex the communication, the more profound the problem will be. Attempt to hold runner communications to the transmittal of no more than three new items or changes in strategy or a combination of both.

The best use of a runner is to select a trained manager or subordinate who thinks like you do. In this manner, the runner can add that little touch that may be necessary to get the point across. Also a trained manager can more readily transmit feedback to you, because the runner has the ability to tie together loose ends through personal observation.

Written or Text Messages as a Means of Communication

Although written messages are seldom used, they can be an excellent communications tool. Diagrams, maps, sketches, etc. can provide more information than any other means of communication. Written directions limited to one task delivered by a runner can be extremely useful.

The advent of mobile data terminals, palm held or laptop computers has opened a whole new means of communications on emergencies. Although costs have held most agencies back from adopting this technology in the future it is bound to open another means of communications that will improve the manner in which we operate. In addition to sending text messages to key officers the day may not be that far away when heads up displays will be available in self-contained breathing apparatus.

Face-To-Face as A Means of Communicating

Some agencies rely heavily on other forms of communication, yet any other form of communication can not replace the best: face-to-face communications. Any problems associated with face-to-face communications can be overcome on an emergency with adequate training.

Training involves the setting of simple guidelines before an emergency. For example, have all supervisors report of the Incident Commander unless advised to do otherwise. Instruct subordinate managers who are within line of sight or a reasonable distance to report in person. Establish ground rules for the use of other means of communication. It is important that supervisors recognize that when orders are unclear, an objective is uncertain, actions to be taken are not identified clearly, or where safety is a real concern, that face-to-face communication must be utilized.

Generally, a subordinate manager should be briefed on a face-to- face basis if authority and responsibility is being released. The plan should be explained, organization outlined, diagrams provided, controls specified, and reference made to what can be seen from the command post.

Performance-oriented communications is what emergency managers emphasize. Face-to-face communications is the most effective method to release authority and responsibility direct activities or explain coordination. A network that uses other methods exclusively does not communicate all vital and individual meaningful information that will cause performance problems.

FEEDBACK SYSTEMS

Feedback systems are designed to provide an incident manager and supervisors with pertinent information concerning the management of the emergency. Feedback is a process in which the factors that produce a result are then modified, corrected, strengthened, etc. by inter-communication between managers, supervisors, and subordinates. Depending on how feedback is used on an emergency it can be an advantage or disadvantage. Feedback has to focus on the strategical and tactical objectives, as well as the timeframes established by the Incident Commander.

Feedback has to follow some type of radio protocol or discipline, otherwise the amount of unnecessary confusion and contradictory information can cause a total breakdown in meaningful communication. Supervisors and managers must learn and teach subordinates that only information that will have an impact on the established plan has to be feedback through the organization. There are three types of feedback systems: negative, positive, and controlled.

Negative Feedback

Negative feedback is based on the principle of management by exception. When using this method of feedback the manager or supervisor only wishes to hear when something cannot be accomplished. This type of feedback presumes that everything is going as planned. This method is very effective as long as the person in charge can see all the activities and can make any necessary adjustments before a problem has major impact. After an initial communication or briefing, no future communication is necessary until a problem arises or the job is completed. When an operation is spread over a large geographical area where direct line-of-sight is not possible, the ability to control activities is weakened because the supervisor may be forced to re-adjust objectives if certain operations are not successful. These adjustments should be made when first detected, not at a point when an objective is no longer attainable.

Positive Feedback

A positive feedback system requires much more feedback than a negative system, although It does provide a much better picture of what is transpiring in the different operational areas. The difference between a negative and positive system is that the manager or supervisor with the positive system is advised when each portion of a total operation is completed. The theory behind this system is that if managers and supervisors are aware of what has been accomplished, better control will be exercised.

However, when a large or complex emergency occurs, the amount of feedback received can become overwhelming. Unlike the negative system where too little information may be received, too much information is usually received in too rapid a manner. With either system, managers and supervisors can become unable to produce reasonable deductions and/or decisions.

Controlled Feedback

If used properly controlled feedback is a method of managing feedback and activities at the same time. It can be positive or negative depending on the need. Controlled feedback is used to teach, in addition to supporting the management philosophy of the department. The principle is to reinforce prior communications, redirecting or strengthening a supervisor's or subordinate's role in the total operation. Basically, controlled feedback requires a manager or supervisor to seek or defer information as necessary to maintain adequate awareness of activities in order for the organization to reach the stated objectives.

Controlled feedback works on a principle of expanding roles. For example: a subordinate reports an occurrence that is not part of the original plan. The manager reacts to the feedback, "I copy, can you handle?" In this case, the person in charge has managed or controlled the feedback by placing more authority on the subordinate. The subordinate now has two choices: to report that they can or cannot handle. If the feedback at this point says, "I can," then the next question the supervisor may want to ask is, "What additional resources do you require?" Should the response to the question be "I cannot handle," then the manager may respond with "What do you need?" or "What do you suggest?" In each situation, the decision has been shifted to the subordinate who is in the best position to take or recommend the correct action. In this manner, the manager has controlled the feedback, at the same time widened the scope or reinforced the subordinate's responsibility and authority with a minimum amount of communication. If managers and supervisors follow this approach time after time on several different situations, subordinates will become more attuned to handling problems and advising you as an action starts or suggesting a course of action to meet a changing situation. This in turn will further reduce the amount of feedback that improves communication quality and content.

Seldom does an emergency manager have to defer information if personnel are adequately trained in feedback procedures. The best we can do is learn when it is necessary to control feedback and how to handle a situation tactfully with the least amount of confusion and communication. Minor problems should be addressed after an emergency has concluded, usually on a one-to-one basis. Major distractions may have to be dealt with during the incident. The best way to handle feedback problems is to recall the individual to the command post or send someone to re-establish authority and responsibility, redefine the plan with timeframes, or provide exact direction on the type of required feedback.

COMMUNICATION RULES TO FOLLOW

Good communications starts with the Incident Commander. If Incident Commander concentrates on the strategy, sets a communications example and follows the rules, such as the eight below, ICS Strategy will improve communications on emergencies.

1. When releasing authority or responsibility to a subordinate, communicate face-to-face not over the radio. This action may add two or three minutes to your initial operation, but the translation to action will most likely be correct thus saving much more time than invested.

2. Plan your radio communications. Think about not only "what you are going to say," but also "how it may be interpreted."

3. Hold radio communications to one action or task. When communicating by radio, the chance of a second order being correct is 50/50. Generally, if you find more than one "and" in a communication, it translates into two jobs.

4. Use note pads to keep track of assignments and tasks. Insist that subordinates do the same if they are given authority and responsibility for an operational area.

5. Have radio communications repeated back to you before they are translated into action.

6. Whenever possible, position yourself in a location where a line-of-sight communication can be maintained over critical operations.

7. Use the radio to check progress when line-of-sight cannot be maintained. This should be done in a manner that requires more than an "Okay" response. For example: Is the rear of the building laddered?" Don't ask, "How are you doing?"

8. If a communication is unclear from a subordinate, clarify by asking for a repeat, having the subordinate return to your location, or making an inspection yourself.

One other procedure that is important to good ICS communications is the concept of expanded/reinforced authority. When subordinate officers bring up problems follow these steps, in order:

1. Ask them if the strategic objective can still be met. This reinforces the fact that if items do not effect the strategical objective then they do not have to be communicated.

2. Ask them if they can handle the problem and what impact it will have on the objective.

3. If they cannot handle the problem, what do they need to handle it or what do they recommend.

Finally, we must never presume that others know what we are talking about or that we are sure of what others are talking about. We must be explicit, get feedback and ask questions to ensure our spoken communication is fully understood. Remember the ability to communicate is not given to us, we must learn it; and after we have learned it, we must continue to practice it or the skill will be lost.

NOTES:

NOTES:

CHAPTER VI

ANALYZING and FINAL THOUGHTS

ANALYZING FOR PROGRESS

Analyzing can be broken down into two different elements: one is during the emergency; the other is after the emergency is over. One of the strongest aspects of ICS Strategy is if you follow the steps of Planning, Organizing, Communicating and Controlling, the fifth element of Analyzing for progress, or the lack thereof, during the emergency becomes self-fulfilling. This is because the first four elements have provided you with the foundation to know what it transpiring (good or bad) at all times.

Yet, it is still a wise Incident Commander who will at some point in the operation, review everything that is occurring or is about to occur. A good rule of thumb for telling you that you need to take a more objective look at what is happening is if your command area is hectic. If you have done everything we have suggested in the first four elements you should be experiencing something known as the loneliness of command. This is a feeling that you should be doing something. The real truth at that moment is that you have done what is necessary. If you do not get this feeling, start back through the first four elements to place your operation back on track.

ANALYZING THE RESULTS

The most difficult part of using any emergency management system is analyzing the performance of personnel in using the system after the incident. How many of us have attended an incident analysis, sometimes called a critique, post-incident analysis or critical incident analysis and come away with the feeling nothing was accomplished? Unfortunately, many times we do not seek to correct problems, but to praise ourselves on good performance or criticize others. The emphasis of any after-the-incident review should be to seek improvements at the strategical and tactical level. In order to accomplish this most officers and departments will have to change their approaches to the incident analysis.

The two most prevalent types of incident analysis are termed show-and-tell and blame-finding. Oftentimes we spend so mush time explaining what happened or defending our actions that we forget the purpose is to improve for tomorrow.

I have never been on an incident of any consequence as an Incident Commander or subordinate officer that was without at least one problem; and it is doubtful, if we are honest, if anyone in this business has been or ever will be. That is a fairly significant statement, however true, and one that we must deal with. It is with that thought that we must approach the aftermath of each incident. This is not to say that we have not been on some well-run emergencies; yet, there always seems to be at least one area where something could have been done more efficiently.

The most difficult part of incident analysis is remaining objective. Objectivity should be the number one issue. It is for this reason that sometimes it is better to have a person not involved with the incident conduct the incident analysis. An objective incident analysis identifies what occurred and why, seeks to correct problems and find reasonable solutions.

The most important - and most difficult - thing that officers can do to improve an incident analysis is admitting their own errors. It is very important for the Incident Commander to start the process by admitting personal errors.

Types of Incident Analysis

The three basic types of post-incident analysis are: informal, formal and comparison. Each of these methods has its advantages and disadvantages; and depending on the situation, each may have its place in an organization using ICS Strategy.

Informal Analysis

The best place for an informal analysis is in a relaxed atmosphere. However, we have seen effective informal analysis conducted at the scene of the emergency after the situation is under control. To be truly effective, the Incident Commander should lead a general discussion starting with admission of the problems the commander may have caused or recognized especially as they relate to the Plan and ICS Organization. There should be no placing of blame except on oneself. The message the Incident Commander should convey is, "It's okay to say I made a mistake."

Informal incident analysis is often called coffee-table strategy and tactics. One of the real advantages of the informal analysis besides the relaxed atmosphere is there is usually no documentation necessary. However, this is also its greatest disadvantage because other officers often do not hear about problems and how to correct them to improve future operations of the same nature.

Formal Analysis

A formal analysis means documentation. Here writing about findings becomes important. We often suggest that formal analysis be limited to incidents of a significant nature. Because the definition of a significant incident can vary greatly from agency to agency, it suffices to say that any incident involving injury or death to emergency personnel must be considered significant.

The advantages of a formal incident analysis are that they usually involve all personnel and they are documented. The disadvantages are delays in scheduling (especially if multiple agencies are involved), lack of objectivity, and sometimes a lot of time is spent on matters of little importance for future improvement. They also can be unruly or hard to control; they can turn into finger pointing and back slapping; and they can be very time consuming. Because of delays in conducting this type of analysis, it is recommended that key officers write down any issues they observe as soon as possible after the incident.

The best way to avoid many of the problems of a formal incident analysis is to have an impartial facilitator and a recorder. The facilitator should set a time and place shortly after the incident, usually not to exceed a week. A specific time period should be established with a starting and ending time. An agenda should be prepared and distributed prior to the meeting to all personnel who will be attending.

The agenda should list some of the known problems, ask individuals to come to the incident analysis and be ready to identify others, and finally state that at a specific time the group will discuss possible solutions.

After the formal analysis is completed, the recorder should file a report for distribution throughout the agency and/or involved agencies. The report should identify the problems without reference to anyone or any group, and should list the possible solutions.

Comparative Analysis

The comparative analysis differs from the formal in that only key management personnel are involved. Again, it is often best if an impartial facilitator conducts the analysis. This analysis should also be conducted within one week of the incident and be limited in time. The facilitator starts the comparative analysis by asking each officer in a key position to write down the perception of the "plan and organization" as it was related or observed by them. The facilitator then compares these usually differing forms, and possible problems listed on a chalkboard, overhead transparency, etc. The personnel are then asked to identify the problems that did occur and to provide others that they believe were not listed; then a discussion occurs regarding these problems. The facilitator notes any differences of opinion, and attempts to refer these back to the Plan and ICS Organization. Generally after this occurs solutions are discussed and the session is concluded.

The facilitator then files a written incident analysis report complete with the different perceptions by the various officers of the Plan and ICS Organization and list of the identified problems and possible solutions. This report is then sent to all involved companies, as well as other officers and companies who were not part of the operation.

We have attempted to show you how the incident management cycle can work for you. Only through objective analysis of today's performance can we improve for tomorrow. We all need to reflect on what has occurred and to seek better ways for the next item. If we fail in this area, we fail other personnel, the department, and ourselves. In short, we have gained nothing.

FINAL THOUGHTS

The face of today's emergency fields are changing and perhaps no place is this more evident than in the area of command. ICS has impacted managers so much that a new set of challenges has emerged. Are you a strategist, tactician or practitioner?

ICS has begun or affects all the emergency fields. As it continues to evolve more improvements will occur. Each innovation will bring about improvement in our abilities to work with one another. But ICS will not begin to solve all our problems until planning becomes a part of the total management process. We must not lose sight of the fact that ICS is nothing more than an organizational system that is a tool to be used by the Incident Commander in any manner that is appropriate based on the situation.

Many officers are so intent getting things organized that they start looking for and appointing people to ICS positions before considering other aspects of the emergency. They are quick to point out that an Incident Commander should be a strategist and a large part of the job is organization. However, in many cases, they have lost sight of the fact that the primary objective of an Incident Commander is to manage an emergency. Managing means more than organizing, one cannot organize in a vacuum; there must be a purpose. This means planning comes first; followed by communicating, organizing, controlling and evaluating.

The problem is twofold: (1) officers must not use organization as the first step in managing an emergency; and (2) they must be taught to plan. This means that a realistic objective(s) must be set; time frames considered to measure the progress or the lack thereof; and a logical progression of events provided to attain results. If this is not done, all the organizing in the world will not prevent an Incident Commander from making large errors.

What makes a good Incident Commander? Many schools of thought abound this question. I have taught for years three principles. An Incident Commander must:

1. Be a tactician. Not so much that he/she knows how to operate every piece of apparatus and equipment, not to the extent that every job or task is guided by direct communication. But an incident commander should know the practical realities as to the capabilities of the resources, including the people, are known in relation to the situation.

2. Know about the behavior of the element causing the emergency.

3. Be a strategist - have the ability to: (a) plan; (b) communicate; (c) organize, (d) control; and (e) analyze for the duration of the incident. To say the least, none of these things can be done without assistance in every case; but the Incident Commander must be capable of all these from the onset of the problem.

My recipe for an outstanding Incident Commander: Take strategy and tactics; add a person who is a practitioner of both; mix with knowledge of the elements involved; and simmer with time and experience.

The Incident Commander for today and tomorrow must be a practitioner of all aspects encompassing his/her responsibilities. One cannot be a strategist without being a tactician and vice versa. Officers must train to be balanced strategists, not just organizers. After all, ICS or organization by any other name is not the total answer.

NOTES: